Entreprise Interoperability

Entreprise Interoperability Set

coordinated by
Bernard Archimède and Jean-Paul Bourrières

Volume 1

Entreprise Interoperability

INTEROP-PGSO Vision

Edited by

Bernard Archimède
Bruno Vallespir

iSTE

WILEY

First published 2017 in Great Britain and the United States by ISTE Ltd and John Wiley & Sons, Inc.

ISTE Ltd
27-37 St George's Road
London SW19 4EU
UK

www.iste.co.uk

John Wiley & Sons, Inc.
111 River Street
Hoboken, NJ 07030
USA

www.wiley.com

Library of Congress Control Number: 2017932961

British Library Cataloguing-in-Publication Data
A CIP record for this book is available from the British Library
ISBN 978-1-78630-084-3

Contents

Foreword

I am glad that the INTEROP-VLab Grand-Sud-Ouest (GSO) pole, which is well known by the international interoperability community for its commitment to advancing the frontiers of interoperability in crisis management, e-Health and Transport, has decided to publish an annual series on Enterprise Interoperability of which the present book is the first opus. And I am grateful for having been invited to write this foreword: this is an honor for the Head of Unit at the European Commission who had the privilege to witness and accompany the development of the founding project – the INTEROP Network of Excellence – and the progress of its remarkable spin-off – the International Virtual Lab for Enterprise Interoperability.

When interoperability became a topic of extensive discussion in the late 1990s, the term was fundamentally referring to the ability to exchange functionality and interpretable data between two software entities. Over the past fifteen years much work has been done on interoperability of applications and software, in particular within Europe that can legitimately claim pride and honor for its achievements. This has been the work of a generation, which will have an enormous impact for years and even decades to come, on the ability to make systems talk to each other.

However, my feeling is that interoperability is the epitome of an unending race between disruptive technological change and relentless modeling and standardization efforts. I will give just two examples.

We know that the future of manufacturing lies in *cybermanufacturing* (i.e. research in and prototyping of operating systems and applications to create platforms for large-scale manufacturing), *biomanufacturing* (i.e. basic research at the intersection betweem biology and engineering to revolutionize healthcare) and *nanomanufacturing* (i.e. basic research at the intersection between chemistry, physics and engineering to manipulate matter at atomic and molecular scales).

We also know that the future of the Internet of Things – empowering trillions of devices (or "Things") to tell us how best to use them – lies in a common platform that works the way the smartphone does today, which requires, in particular, semantic interoperability between architectures, standards and ontologies by design as well as security by design.

The future of manufacturing and the Internet of Things are for me the biggest challenges facing the interoperability community. I am fully aware of how unfair it may look to say that work on interoperability has just begun. So much has been done so well already. Perhaps there are words like interoperability that have a kind of destiny. This is the magic and the tragedy of such words – they have roots that can grow very long but they also have nodes that make it sometimes hard for them to take off and fly where we want them to. I experienced this feeling during the 8 years in the European Commission while managing the eBusiness Unit (which after some years was renamed the Networked Enterprise and Radio Frequency Identification Unit).

I was appointed Head of the eBusiness Unit of European Commission's Directorate-General Information Society and Media (DG INFSO-D/5) on 1 March 2004 during an incredibly interesting and rich period of the European Research Area (integrating, strengthening, structuring). With the Lisbon Strategy and its eEurope action plan (the EU to become the most dynamic and most competitive knowledge-based economy within 10 years), the Barcelona Declaration (education, employment, enlargement) and the Gothenburg objective (sustainability), I inherited a huge and diverse portfolio of projects: 12 Framework Programme 6 (FP6) IST[1] projects, 177 Framework Programme 5 (FP5) IST projects and 16 Framework Programme 4 (FP4) ESPRIT projects (whose contracts were not yet closed). The 12 FP6-IST projects were retained by European Commission Decision following the evaluation of the proposals submitted to Call 1 of Strategic Objective 2.3.1.9 "Networked Businesses and Governments". These projects were to be complemented soon by other projects allocated to my unit from FP6-IST Call 2 "Applications and Services for the Mobile User and Worker" (Strategic Objective 2.3.2.6, 2 projects) and the Joint Call between the IST Programme and the Nanotechnology and Nanosciences, Knowledge-based Multifunctional Materials and New Production Processes and Devices (NMP) Programme (Strategic Objective 2.3.3.1, 4 projects).

The proliferation of projects selected from a succession of calls and addressing similar technologies, applications, services and related socio-economic issues, had grim prospects in terms of coherence, effectiveness and impact. Therefore, I decided

1 IST: the *Information Society Technologies* programme (under Framework Programmes 5 and 6).

to organize consultations as a key process between projects. Convinced that it was possible to make the value of the work in this field exceed the value of the individual projects, I invited people and their organizations involved in the new FP6-IST projects to come together and to share their knowledge and experience to their mutual benefit. The main objectives of the consultations were the following: to maximize technological relevance for mid- to long-term R&D work on ICT for Business and support system integration and engineering in specific application areas of strategic interest; to assess socio- and techno-economic factors, regional variations and trends, as well as developments in policy and regulation, in order to interpret their impact on specific R&D actions; to foster pre-normative research, interoperability, benchmarking and best practice; to identify critical success factors for trials and pilot demonstrations; to integrate assessment and technical validation results to ensure that valid results of projects were properly taken into account in the ongoing work of others; to create synergy with Member State activities (in the framework of the European Research Area) and other European programs (Eureka); and finally to stimulate and enhance collaboration on common communication, dissemination and exploitation aspects.

During the Spring of 2004 my unit created four "clusters" of projects to manage consultations, including one on Enterprise Interoperability, initially composed of four projects: ATHENA (integrated project), INTEROP (network of excellence), NO-REST (specific targeted research project) and TRUSTCOM (integrated project). This was the beginning of an 8-year long period of extensive collaboration between EU-funded projects in the field of Enterprise Interoperability (2004–2012) with the work constantly stimulated and assessed by my unit through the regular running of consultation meetings in Brussels.

For the sake of completeness, I need to stress that European interest in Enterprise Interoperability started before I headed the eBusiness Unit and created the cluster. In fact, the work resulted from a road-mapping initiative called IDEAS (in fact a thematic network named Interoperability Developments of Enterprise Applications and Software), launched by DG INFSO in 2000 as a response to repeated requests by European Industry (e.g., Aerospace, Automotive, Manufacturing) to address the issue of the cost of non-interoperability of information technology solutions. Joël Bacquet, Project Officer in the eBusiness Unit, then managed by Jesus Villasante, had initially invited Guy Doumeingts, Professor at University of Bordeaux, to create a Task Force on Enterprise Interoperability in order to assess key issues and report to the Unit research options for tackling them. After two years of extensive discussions, the Task Force composed of fifteen experts completed a review of the state of the art on Enterprise Interoperability and proposed the IDEAS roadmap. I want to express here my gratitude to Joël Bacquet and Guy Doumeingts, two pioneers, for their extraordinary dedication to the European cause, their prolific scientific work, their unfailing drive to tackle problems until they are solved, and of course for their high ability in carrying this endeavor through at a time of great confusion and uncertainty.

Indeed, it was a challenge in 2004 to bring together the various individuals and communities involved in Enterprise Interoperability. The four projects of the cluster, mainly ATHENA[2] (industry-driven) and INTEROP[3] (academically-driven), had different approaches to the work. Furthermore, several frameworks emerged during the same period, among which: the UK's e-Government Interoperability Framework enabling the seamless flow of information across government/public service organizations (e-GIF, 2000); the European Interoperability Framework developed under the IDABC (now ISA) program (EIF, 2004); the Enterprise Interoperability Centre, originated from ATHENA and established as a not-for-profit legal entity under the Belgian jurisdiction (EIC, 2006); the ATHENA Interoperability Framework for enterprise applications and software systems (AIF, 2007). The Enterprise Interoperability cluster, superbly coordinated in my unit by Arian Zwegers (2004–2005) and Cristina Martinez (2006–2012), and which changed name twice[4], played a catalytic role in furthering concertation and consensus thanks to a portfolio of about 30 FP6-IST and FP7-ICT projects by mid-July 2012, i.e. when DG INFSO became DG CONNECT.

The International Virtual Laboratory for Enterprise Interoperability, which celebrates its 10th Anniversary at the same time as this book is published, represents to me the only example of a spin-off of a network of excellence that has exceeded all success standards set by the European Commission thanks to the commendable commitment of its team to building and maintaining over the years a world-class organization with a dynamic corporate culture. When I attended an I-VLab meeting on 6 April 2009, I pronounced a few words which 8 years later resound in the same way:

"The I-VLab initiative reflects the commitment of the networked regional poles to accept and promote the new spirit that has taken hold in Europe in the wake of the INTEROP Network of Excellence: a new sense of community, a new willingness and pledge to develop common strategies for advancing research in Enterprise Interoperability, a new understanding that the public sector and the private sector are

2 ATHENA: Advanced technologies for interoperability of heterogeneous enterprise networks and their applications; from 1/02/2004 to 31/03/2007; 25 partners; coordination: SAP AG.
3 INTEROP: Interoperability research for networked enterprises applications and software; from 1/11/2003 to 30/04/2007; 51 partners; coordination: University of Bordeaux I Sciences and Technologies.
4 The cluster was extended a first time in 2007 to include also Enterprise Collaboration (Enterprise Interoperability and Collaboration, EI&C) and a second time in 2009 to include Digital Ecosystems (Future Internet Enterprise Systems, FInES). In 2014, the FInES community agreed to rename the cluster Digital Business Innovation to place it at the confluence of Digital Innovation, Entrepreneurship and Technology Foundation.

all in this together, also a new recognition of the catalytic role of the European Commission, and finally a new readiness on the part of the actors to take a multidisciplinary approach by merging the three areas of Architecture & Platforms, Enterprise Modelling and Ontology".

Looking towards the future, the challenge is obvious. In Europe, we will have to take many important steps on the long path to interoperability, by adopting resolutely and systematically a collaborative, co-creative and multidisciplinary approach. Interoperability solutions are primarily a matter of engineering, no doubt, but I learnt from mathematicians like Michael Leyton (Rutgers University, New Jersey) and research scientists like Ted Goranson (Sirius-Beta) that it is possible to contemplate work on the foundations of Interoperability as a science. We have unprecedented opportunities to improve our understanding of interoperability for the benefit of companies and society. Breathtaking advances in interoperability can take place at breathtaking speeds as long as we realize that the future in the digital world is less and less incremental and more and more disruptive. This is an essential prerequisite for tackling the two big challenges that are the future of manufacturing and the Internet of Things, among others which I already mentioned above.

In the pages that follow the authors have sought to offer a both theoretical and operational framework capable of providing a synthesis of current research programs in Enterprise Interoperability. The holistic, systemic, integrated approach which they have followed is an invitation to shift from thinking in terms of separation or grounding one issue in another, to thinking in terms of entanglements, thus allowing to maintain the irreducibility, heterogeneity and autonomy of various types of entities while exploring how they influence one another. This approach could have been extremely difficult to follow, but very wisely and intelligently the authors have written a book that is accessible to a wide audience, without compromising the force of the intellectual analyses, and that can be utilized by others as rich and unique material in a variety of disciplines and practices, generating new questions, new ideas and new projects. This book, and no doubt those that will follow, are great accomplishments to serve the cause of Enterprise Interoperability. The fact that this initiative is born in Europe is for me another source of contentment and pride.

Gérald SANTUCCI
Directorate-General for Communications Networks,
Content and Technology
Adviser for Cross-cutting Policy and Research Issues
European Commission

Introduction

The rise of the notion of enterprise interoperability can be traced back to the beginning of the 2000s, when large companies faced difficult challenges when attempting to limit the costs of computing and information management solutions and adapting the IT tool to novel needs stemming from the two following trends. The first trend is related to the demand for de-compartmentalization. This tendency can apply either internally by requiring that the various company departments collaborate to avoid developing silo management patterns, in an approach popularized by Business Process Management, or externally through the creation of ever larger networks of companies, the most typical example of which being supply chains. The second trend is due to accelerating changes at all levels, from the strategic level (takeovers, company partnerships) to, for example, information technology advances or product innovation. The larger European groups have raised awareness so much so that the European Commission has introduced the issue of interoperability in the IST program (5th Framework Program) and financed several projects aimed at making a body of knowledge about this domain. In this introductory chapter, we present a short historical account of the emergence of enterprise interoperability, which has now become an established multi-disciplinary research theme with high industrial stakes.

I.1. Initializing enterprise interoperability research by EU-funded projects

European research on enterprise interoperability was initially based on three pioneer projects of the IST program (User-friendly Information Society): IDEAS-TN (FP5), INTEROP-NoE (FP6) and ATHENA-IP (FP6). The determining leverage effect of three key subsidized projects during 2003–2007 resulted in the maturation

Introduction written by Bernard ARCHIMÈDE, Jean-Paul BOURRIÈRES, Guy DOUMEINGTS and Bruno VALLESPIR.

of enterprise interoperability knowledge and its deployment toward both researchers and practitioners as well as in the dissemination of scientific results through the biannual international conference I-ESA. Finally, as expected by the European Commission, from the initialization phase through to funded projects, the creation of INTEROP-VLab, a durable and autonomous organization to consolidate and deploy a networked research on Enterprise Interoperability (Figure I.1).

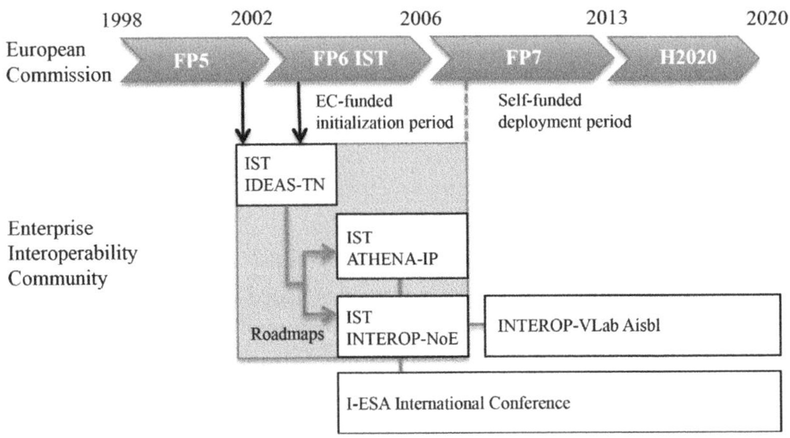

Figure I.1. *From subsided European projects to autonomous INTEROP-VLab*

The main steps of the enterprise interoperability development process in Europe are briefly presented below.

I.1.1. *IDEAS*

The Thematic Network IDEAS (Interoperability Developments of Enterprise Applications and Software-Roadmaps) has elaborated a strategic roadmap in the domain of enterprise application and software interoperability for the next 10 years. Acting as a working group for the European Commission, this project has proposed guidelines intended for the FP6 work program to develop a multi-disciplinary research work on Interoperability in Europe. The research roadmap elaborated by IDEAS recommended addressing interoperability issues in the enterprise *not only at the IT level*, but also to cope with the technical communication between people involved in intra-enterprise organization as well as in enterprise ecosystems. IDEAS has emphasized the necessity of deploying a multi-disciplinary approach to enterprise interoperability based on the integration of three disciplinary components: (1) Ontologies to identify the interoperability semantics within the enterprise and between networked enterprises, (2) Enterprise Modeling to identify Business Processes requirements for interoperability and (3) Architectures and Platforms to enable and implement IT solutions for interoperability (Figure I.2).

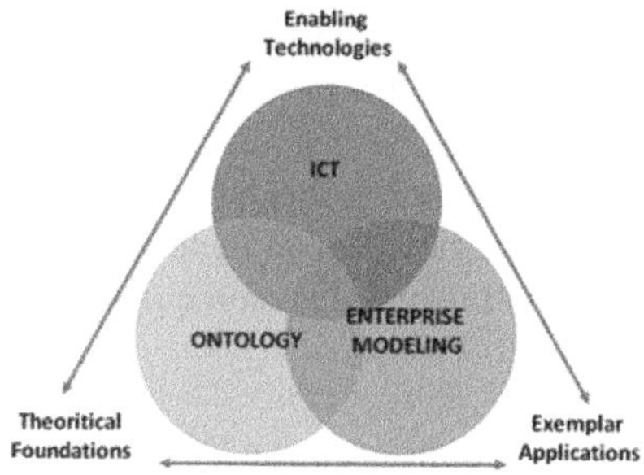

Figure I.2. *Three disciplinary pillars of enterprise interoperability (source: IDEAS Project)*

IDEAS (IST-2001-37368) **Title**: Interoperability developments of enterprise applications and software-Roadmaps **Start date:** 1 June 2002 **Duration:** 12 months **EU Funding:** 400 k€	**Participants:** UNIVERSITE BORDEAUX 1 (FR) Coord. AIDIMA (ES) BAAN DEVELOPMENT B.V. (NL) CENTRO RICERCHE FIAT (IT) COMPUTAS HOLDING AS (NO) EADS CCR (FR) GRAISOFT (FR) GRUPPO FORMULA S.P.A (IT) IC FOCUS LIMITED (UK) INTRACOM S.A. (GR) SAP A.G. (DE) UNINOVA (PO)

I.1.2. *INTEROP-NoE*

Following IDEAS's roadmap, the topic "Interoperability for enterprise applications" appeared explicitly in the 6th Framework Program of the EC. At that time, the Commission began to deploy the ERA (European Research Area) initiative to structure the European research resources in emerging fields. In particular, the purpose of the new instrument, NoE (Networks of Excellence), was to help key research players, both from academia and industry, to establish durable thematic European communities.

In this framework INTEROP-NoE (Network of Excellence in Interoperability Research for Networked Enterprise Applications and Software, FP6 508011) was funded with the mission of integrating and animating a community in enterprise interoperability and, beyond, establishing a sustainable organization able to continue networked research activities in the field after the EC-funded period.

Orchestrating a large consortium of over 50 participants from over 15 member states, INTEROP-NoE in its 42 months of intense joint research work has strongly contributed to the integration of the three components of knowledge (Figure I.2) recommended by IDEAS-TN to form a new multi-disciplinary European community dedicated to enterprise interoperability, with a very significant critical mass. Simultaneously, an important effort has been made to disseminate the concept of enterprise interoperability and the attached methods and tools to have an impact on national research bodies and higher education curricula as well as on vocational training and standardization.

Last but not least, it must be emphasized that INTEROP-NoE, a research initiative, has be run in collaboration with a second project, ATHENA-IP, focusing on the industrial issues of Interoperability and innovation in IT. The synergy between INTEROP-NoE and ATHENA-IP was ensured by a large overlap of both consortia.

INTEROP-NoE (IST-2003- 508011)

Title: Interoperability research for networked enterprises applications and software
Start date: 1 November 2003
Duration: 42 months
EU Funding: 6.5 M€

Participants:
UNIVERSITE BORDEAUX 1 (FR) Coord.
AIDIMA (ES)
BOC INFORMATION SYSTEMS GMBH (AU)
UNIVERSITAET OLDENBURG (DE)
CENTRE DE RECHERCHE HENRI TUDOR (LU)
CNRS (FR)
CENTRO RICERCHE FIAT (IT)
CIMMEDIA Ltd (UK)
CIMOSA ASSOCIATION E.V. (DE)
CONSIGLIO NAZIONALE DELLE RICERCHE (IT)
DFKI GMBH (DE)
ECOLE CENTRALE DE LILLE (FR)
ECOLE CENTRALE DE NANTES (FR)
FACULTES UNIVERSITAIRES NOTRE-DAME DE LA PAIX ASBL (BE)
FRAUNHOFER GESELLSCHAFT (DE)
GRUPPO FORMULA S.P.A. (IT)
HELSINGIN YLIOPISTO (FI)
HOGSKOLAN I SKOVDE (SE)
INRIA (FR)
INSTITUT NATIONAL POLYTECHNIQUE DE LORRAINE (FR)
ITREC – GESTION (FR)
KUNGLIGA TEKNISKA HOGSKOLAN (SE)
LIBERA UNIVERSITA DI BOLZANO (IT)
LUISS LIBERA UNIVERSITA INTERNAZIONALE DEGLI STUDI SOCIALI GUIDO CARLI (IT)
NORGES TEKNISK-NATUR VITENSKAPELIGE UNIVERSITET (NO)
POLITECNICO DI BARI (IT)
POLITECNICO DI MILANO (IT)
SINGULAR (GR)
SINTEF (NO)
STICHTING KATHOLIEKE UNIVERSITEIT BRABANT (NL)
STICHTING TELEMATICA INSTITUUT (NL)
TECHNISCHE UNIVERSITEIT EINDHOVEN (NL)
TROUX TECHNOLOGIES AS (NO)
TXT E-SOLUTIONS SPA (IT)
UNINOVA (PO)
UNIVERSIDAD POLITECNICA DE VALENCIA (ES)
UNIVERSITA DEGLI STUDI DI BRESCIA (IT)
UNIVERSITA DEGLI STUDI DI ROMA "LA SAPIENZA" (IT)
UNIVERSITA DEGLI STUDI DI TORINO (IT)
UNIVERSITA POLITECNICA DELLE MARCHE (IT)
UNIVERSITAET DUISBURG-ESSEN (DE)
UNIVERSITAET KLAGENFURT (AU)
UNIVERSITAT JAUME I DE CASTELLON (ES)
UNIVERSITE DE GENEVE (SZ)
UNIVERSITE DE LAUSANNE (SZ)
UNIVERSITE DE NANTES (FR)
UNIVERSITE HENRI POINCARE NANCY I (FR)
UNIVERSITE NANCY II (FR)
UNIVERSITETET I BERGEN (NO)
UNIVERSITY OF KENT AT CANTERBURY (UK)
UNIVERSITÀ DEGLI STUDI DI MILANO (IT)

I.1.3. *ATHENA – IP*

The objective of ATHENA-IP (Integrated Project on Advanced Technologies for Interoperability of Heterogeneous Enterprise Networks and their Applications) was to be the most comprehensive and systematic European initiative in the field of enterprise application interoperability, removing barriers to the exchange of information in and among organizations. Building on its vision statement that *"By 2010, enterprises will be able to seamlessly interoperate with others"*, ATHENA contributed to enabling interoperability by providing a comprehensive Interoperability Framework.

ATHENA-IP (IST-2004- 507849)

Title: Advanced technologies for interoperability of heterogeneous enterprise networks and their applications
Start date: 1 February 2004
Duration: 38 months
EU Funding: 14.4 M€

Participants:
SAP AG (DE), Coord.
AIDIMA (ES)
AUTOMOTIVE INDUSTRY ACTION GROUP (US)
AVAG HOLDING AG (DE)
UNIVERSITE BORDEAUX 1 (FR)
CAS SOFTWARE AG (DE)
CENTRO RICERCHE FIAT (IT)
CONSIGLIO NAZIONALE DELLE RICERCHE (IT)
DFKI GMBH (DE)
EADS FRANCE (FR)
FACULTY OF ORGANIZATIONAL SCIENCES, UNIVERSITY OF BELGRADE (CS)
FRAUNHOFER GESELLSCHAFT (DE)
FUNDACION EUROPEAN SOFTWARE INSTITUTE (ES)
GRUPPO FORMULA S.P.A. (IT)
IBM UNITED KINGDOM LIMITED (UK)
INSTITUT EUROPEEN D'ADMINISTRATION DES AFFAIRES (FR)
INTRACOM S.A. TELECOM SOLUTIONS (GR)
ITREC – GESTION (FR)
SIEMENS AG (DE)
SINTEF (NO)
SMC INFORMATIONSTECHNOLOGIEN AG (DE)
TROUX TECHNOLOGIES AS (NO)
TXT E-SOLUTIONS SPA (IT)
UNINOVA (PO)
UNIVERSITAT ST GALLEN (SW)

I.1.4. I-*ESA conference*

A joint action of INTEROP-NoE and ATHENA-IP was to launch an international conference dedicated to interoperability and enterprise applications. The I-ESA conference (interoperability for enterprise systems and applications) is a biannual conference, which brings together the world's leading researchers and practitioners in the area of enterprise interoperability: it is a unique forum for the exchange of visions, ideas, research results and industrial experiences, dealing with a wealth of interoperability research subjects for business, enterprise applications and software. The I-ESA conference, nowadays organized by INTEROP-VLaB, will reach its 10th issue in 2018.

I-ESA international Conference on Interoperability for enterprise systems and applications

I-ESA 2005, Geneva (CH)	I-ESA 2010, Coventry (UK)
I-ESA 2006, Bordeaux (FR)	I-ESA 2012, Valencia (ES)
I-ESA 2007, Funchal (PT)	I-ESA 2014, Albi (FR)
I-ESA 2008, Berlin (DE)	I-ESA 2016, Guimaraes (PT)
Special I-ESA 2009, Beijing (CN)	I-ESA 2018, Berlin (DE)

I.2. Consolidation and deployment of enterprise interoperability research: INTEROP-VLab Aisbl

The overall mission of INTEROP-NoE project, ending in 2007, was to find a durable and self-funded organization to exploit the scientific and social heritage outcome from 42 months of joint research in the NoE. Below are briefly presented the international laboratory INTEROP-VLab, created in 2007, and INTEROP-GSO, one of its member organizations implemented in France.

I.2.1. *INTEROP-VLab Aisbl*

The International Virtual Laboratory for Enterprise Interoperability INTEROP-VLab (I-VLab, www.interop-vlab.eu) is a non-profit international association under the Belgian law, which has its headquarters in Brussels. I-VLab acts as a center of excellence in enterprise interoperability. Their activities include the orchestration, the promotion and the dissemination of models, tools and implementation methodologies issued from joint international research in the field of enterprise interoperability and Future Internet.

The structure of I-VLab is distributed and its activity is twofold: at the regional level and at the global level.

The regional entities, the so-called poles of I-VLab, are local member organizations (Figure I.3), both from public research bodies and industry, which perform their own projects and dissemination activities backed onto local policies, incentives and economy. All poles annually report on their activity during the General Assembly of I-VLab.

The global management of I-VLab is achieved at the international level, aiming to capitalize and circulate the knowledge acquired by the regional poles, as well as to elaborate inter-pole activities, such as the elaboration of proposals and the participation in European projects in the field. A scientific committee coordinates the common scientific activities. Thematic activities are performed by specific working groups and task groups to which scientists contribute from various poles. A particular effort is devoted to exploit the results of achieved international projects through community groups. The organization of the biannual I-ESA conference is also under the responsibility of I-VLab's global management.

Since 2008, INTEROP-VLab has been chosen by the European Commission to be the official repository of Enterprise Interoperability (EI) public deliverables, from the 5th, 6th and 7th Framework programs. In the framework of the partnership with the UNITE project, the repository of deliverables has been extended to other topics of the Future Internet, such as Internet of Things (IoT) and Internet of Services (IoS). (See: http://interop-vlab.eu/ei_public_deliverables)

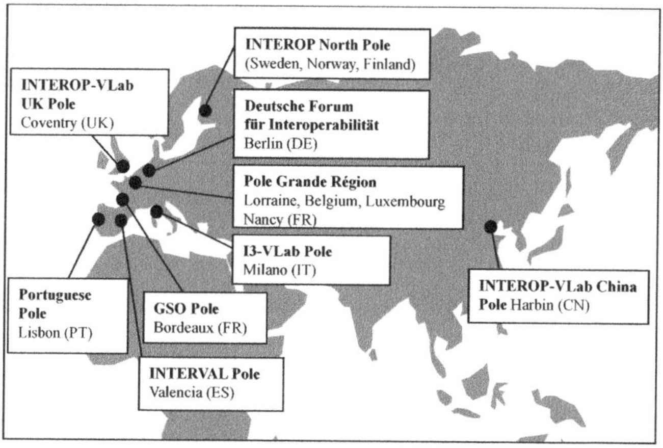

Figure I.3. *INTEROP-VLab network*

I.2.2. *INTEROP-VLab GSO pole*

INTEROP-VLab Grand Sud-Ouest (GSO) is one of the two French poles of INTEROP-VLab. The pole counts 15 academic and industrial member organizations linked by a Scientific Interest Group agreement. The research and project activities of GSO focus on the following three application domains of interoperability:

– crisis management;

– e-Health;

– transport.

The GSO pole organizes the annual JIAE technical days (interoperability and enterprise applications days) during which research results, project achievements as well as case studies and experience feedback are presented.

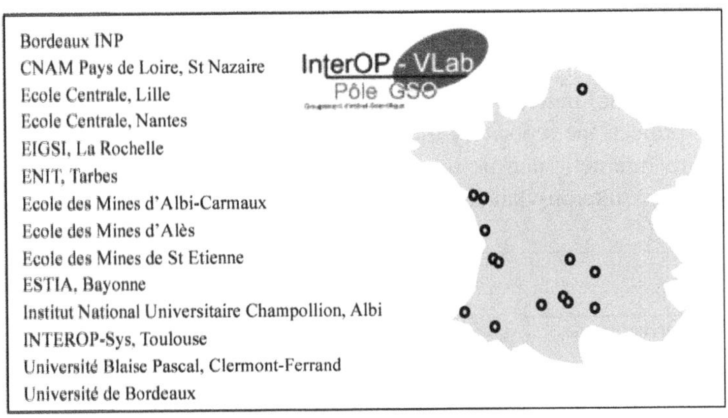

Figure I.4. *INTEROP-VLab GSO pole*

The activity and expertise of GSO pole members are summarized below:

– *Bordeaux Institut National Polytechnique* federates eight public engineer schools. The institute counts 3,200 students in 19 engineering specialties in the fields of Civil Engineering, Health, Natural Resources, Agri-food industry, Energy, Information processing, Transport, Human Resources and Environment. The Institute, jointly with University of Bordeaux and the CNRS, is the supervisory authority of several research laboratories, among which is the IMS Laboratory.

– *Conservatoire National des Arts et Métiers* is a higher-education institution for adults, also responsible for the dissemination of scientific and technical knowledge, technological research and innovation. CNAM Pays de la Loire applies CNAM's strategy in the Pays de la Loire region. For the last 10 years, it has been participating in several research projects in the interoperability domain: French project ISTA3 (3rd Generation Interoperability for Subcontractors in Aeronautics), French project NOSCIFEL (Interoperability research in the transport area) and FP7 European Project MSEE (Interoperability in Manufacturing Service Ecosystem). In these research projects, CNAM of Pays de la Loire has developed vocational training systems to promote project results in industry.

– The Organisation Modelling and Management research group belongs to *Ecole Centrale de Lille*, member of the INTEROP-VLab. This research group works mainly on enterprise modeling, enterprise architecture, information systems and model-driven engineering. It was involved in several collaborative nationally funded projects (ASICOM, ISTA3, etc.), in which the major contribution dealt with model transformation, including model-driven interoperability from both methodological and implementation points of view. This research group is also strongly involved in the MAESTRO chair (Modelling and enterprise ArchitecturE: from Strategy to digital TRansformation of Organisations) of *Ecole Centrale de Lille*.

– *Ecole Centrale de Nantes* is a research and education institution. The main fields of interest and competencies are: Engineering of Industrial Systems/Automatic control/Computer Sciences/Product Design and Manufacturing/Fluid Mechanics, Energetics and Oceanic Engineering/Mechanics, Materials and Civil Engineering/Social sciences: Communication, Foreign Languages and Corporate Culture Department. The IRCCyN laboratory is a Joint Research Unit – UMR 6597 with the *Centre National de la Recherche Scientifique* (CNRS) and its legal representative are *Ecole Centrale de Nantes* (ECN), CNRS, *Université de Nantes* (UN) and *Ecole des Mines de Nantes* (EMN). A former member of FP6 INTEROP-NoE project, the team "Systems Engineering: Product-Performance-Perception" (IS3P) is currently addressing interoperability issues in administrative systems.

– *École d'Ingénieurs en Génie des Systèmes Industriels* is a multi-disciplinary engineering school located in La Rochelle. One research area developed in EIGSI concerns Energy and Coastal protection when the other concerns Integrated Mobility. Research is organized through European projects (INTERREG Atlantic/SUDOE, CIVITAS, H2020, IEE) or national projects (Predit, ANR, ADEME). Research on Integrated Mobility started in 1993 with the first developments of new types of equipment to store energy for electrical vehicle technologies. Considering the local (La Rochelle Urban Community) and regional (Poitou-Charentes) demands and contexts, the research works turned to transport problems linked to the cities and regional strategies on urban and/or land planning and development. These problems generally refer to interoperability issues in information systems and at the governance level.

– The *Laboratoire Génie de Production* (LGP) is a research unit of the *Ecole Nationale d'Ingénieurs de Tarbes*, which has a strong contract research activity involving industrial and national and European institutions. ENIT is associated with the National Polytechnic Institute of Toulouse (INPT) and is part of the Federal University of Toulouse Midi-Pyrénées. LGP is currently involved in several national and European research projects. DIDS (Dynamic Interactions and Decisions for Systems), one of the LGP research teams, is conducting research around several topics, including "Distributed Dynamic Decision" involving the development of new algorithms for solving the problem of collaboration and interoperability in distributed software architectures (data and process integration, as well as semantic interoperability). This research is based on innovative approaches in the field of industrial engineering and ontology-based engineering (domain ontologies, modular ontologies, distributed reasoning on networked ontologies, etc.).

– The Industrial Engineering Center (CGI) is a laboratory of the *Ecole Nationale Supérieure des Mines d'Albi* (member of the Mines-Telecom Institute). CGI is an active member of the INTEROP-VLab. IO (Interoperability of Organizations) is the CGI research team that deals with the collaboration of organizations through advanced decision support systems and interoperability of Information System. Disaster/Crisis Management and Supply Chain Management are the major application fields of these research works. The concerned tools and technologies are: Collaborative Planning, Humanitarian Supply Chain Design and Control, Discrete Event Simulation, Optimization & Operational Research, Performance Assessment, Web-Services, Enterprise Service Bus, Event-Driven Architecture and Complex-Event Processing. The IO team is involved in numerous funded national (ISyCri, SocEDA, GeNePi, etc.) and international (PLAY, DRIVER, C2NET, etc.) research projects on these topics, and has an industrial chair.

– The *Laboratoire de Génie Informatique et d'Ingénierie de Production* (LGI2P) is a laboratory of the *Ecole Nationale Supérieure des Mines d'Alès* (member of the Mines-Télécom Institute). ISOE (Interoperable Systems and Organization Engineering) is a LGI2P research team that develops industry-oriented research on methodological, conceptual and technological approaches for systems engineering (complex, multi-technology and interoperable systems). In this field, ISOE has been, and is currently, involved in national (CARIONER, MAIIEUTIC) and international projects (IMS project PABADIS-PROMISE, Franco-German project ANR-RE(H)STRAIN) to apply its expertise on Model-Based Systems Engineering, properties modeling and Multi-Agents systems principles.

– *Ecole Nationale Supérieure des Mines de Saint Etienne* is a research and educational institution, with six main research fields to meet the current societal and industrial concerns: Energy and Processes/Optimized Materials for aerospace, automotive, nuclear/Industrial Engineering/Environment/Microelectronics/Health. The Research and Education Center of FAYOL Institute is a member of INTEROP-

GSO and develops interdisciplinary research in the field of global and sustainable performance of industrial organizations. FAYOL Institute leads activities on interoperability in the disciplines of computer and manufacturing sciences, with specific application fields in transport systems, health care organizations and servitized manufacturing systems. In these fields, FAYOL Institute participated in national projects (e.g. ANR ServINNOV and FUI Clean Robot) and European initiatives (e.g. CODESNET and OMI-KA2).

– *Ecole Supérieure des Technologies Industrielles Avancées* is a technological school delivering a French engineer diploma. It manages 700 students per year on a trilingual and multi-technological teaching program. The main fields of competences are mechanics, electronics, computing and industrial engineering. Research topics at ESTIA are related to these fields and merged into a single and global project dedicated to the design and prototyping of smart interfaces between Cyber Physical Systems in relation with human beings. Interoperability is a major axis of this research project, which allowed participation in several national R&D projects: ISTA3 on the development of methods and tools for analyzing and improving interoperability between aeronautical companies involved in the co-design of products, and 2PI-MCO on the specifications for building a PLM tool based on the analysis of both human and technological knowledge interoperability.

– *Institut National Universitaire Champollion* is a university located in the north east of Toulouse in the Midi Pyrénées province. The engineering department of INUC is specialized in Computer Sciences and Information Systems for Healthcare systems. Their research domain is e-Health, and they develop activities in the specific areas of user experience and added value of IT. In such a context, interoperability of systems contributing to health care along patient pathways is a subject of interest and a real general requirement from the field of applications. In order to promote their solutions, they have designed and are operating a Living Lab for open innovation in connected health: the Connected Health Lab. It is a platform that can be considered as a test bed for interoperability assessment.

– Created in 2013 by two PhD graduates from *Ecole des Mines d'Albi*, InteropSys is a startup providing online business process management (BPM) application to design, configure, execute and adapt processes throughout their lifetime. All workflows can be directly connected to the IT system using integrated middleware to provide interoperability. This solution is suitable for any businesses of any size, private or public, and does not require any technical skill (zero-code approach). InteropSys is involved in the ANR-GéNéPi project (granularity of the management levels in crisis context) and also invests in internal research to always make the solution easier to use.

– University of Bordeaux, Science and Technology Department, addresses Enterprise Modeling and Interoperability issues as a research and education topic through the IMS laboratory, a Joint Research Unit – UMR 5218 with the CNRS. The former research group, GRAI, consisting of the author of the eponymous enterprise decisional process modeling method, has been leading the key European projects IDEAS-TN and INTEROP-NoE and is one of the contributors to the development of enterprise interoperability as a research topic in Europe. For its expertise in enterprise modeling and performance assessment, the research group has been a member of numerous European projects, in particular ATHENA-IP (FP6) and MSEE-IP (FP7). University of Bordeaux is a co-founder of INTEROP-VLab Aisbl, the International Virtual Laboratory for Enterprise Interoperability, which is currently running.

I.3. Framework and definition of enterprise interoperability

Owing to the renewed focus on their core activity, companies increasingly partner up within networks of companies (supply chains, extended companies, virtual companies), developing partnerships with their suppliers and direct and indirect customers. These partnerships induce changes not only in the organization and internal functioning of the involved companies but also in the collaboration methods, which for a large part rely on information exchanges between, often, heterogeneous company-specific applications. Although the end-to-end control of exchanges of the OSI (Open Systems Interconnection) is extremely reliable, it appeared insufficient to enable companies to always fulfill the needs of the market and improve their efficiency. Several requests to re-transmit information are sometimes necessary to properly understand and then exploit information sent by a partner to a receiving company. This can have negative consequences on the productivity of the latter. To improve the productivity of the companies of a network, we thus need to improve the understanding of exchanged information. This is why interoperability is an issue, which requires specific attention.

Interoperability was considered as the ability of systems or components to exchange information and use it without requiring a specific effort from the users of these systems [IEE 90]. Initially solely considered as a data issue, the notion of interoperability has since grown outside these limiting boundaries to consist of the capacity of peer systems to communicate and have their functionalities accessed [VER 96, IDE 03].

In the case of supply chains that are dominated by large purchasers or OEM (Original Equipment Manufacturer), the issue of interoperability is *de facto* solved because the dominating purchaser dictates its software solutions and working methods to its sub-contractors. This approach to solve interoperability is expensive for the sub-contractors because, in order to not depend on a single purchaser, they need to implement the specific software solution of each of the purchasers with whom they work. Working toward the autonomy of sub-contractors with regard to the independence of their information systems from their purchasers while focusing on the reduction of the costs of concurrently maintaining multiple software solutions, BoostAero Space, an international association, addressed the issue of interoperability. Within, for example, the framework of the SEINE project (*Standards pour l'Entreprise Innovante Numérique Etendue*, standards for the innovative and extended company), they developed interoperability standards and connectors in order to advocate for better practices and ease digital exchanges in the supply chain.

Despite these significant efforts to ease the interoperability of companies' applications, the advocated solutions are still unsatisfactorily implemented because of the numerous conceptual, technological and organizational hurdles that still need to be overcome.

In this context, the ISTA3 project (*Interopérabilité de 3° génération pour les Sous-Traitants de l'Aéronautique*, third generation interoperability for the sub-contractors in the field of aeronautics), which was developed within the framework of the GSO pole, aimed to develop prototypes of customizable, reusable, flexible, adaptive and cheap software solutions for SMBs (small and medium-size businesses) called USI (*Utilitaires de Service en Interopérabilité*, Interoperability Service Utilities). These utilities are more generic than connectors. They apply solutions that overcome various hurdles and rely on concepts and techniques of ontology building and company modeling.

According to GSO, interoperability is reached only if the interaction between two systems can be achieved at the levels of data, resources and business processes with the semantics defined by the business context: *Interoperability can be seen, in a first time, as a problem of compatibility between two systems, not only at ICT level, but at all levels of enterprise (business, process, services, data)* [CHE 03].

Using a different definition of interoperability available in the literature, [CHA 15] proposed a synthesis and considered that interoperability is "*connecting people, data and diverse systems. The term can be defined either technically or comprehensively, in taking into account social, political and organizational factors*". Then, "*two or more devices are said interoperable if, under a given set of conditions, the devices are able to successfully establish, sustain and, if necessary, break a link while maintaining a certain level of performance*". In technical systems, interoperability is "*a property of a product or system, whose interfaces are completely understood, to work with other products or systems, present or future, without any restricted access or implementation*".

I.4. Presentation of the book

The series is devoted to presenting the research works carried out by the organizations belonging to the GSO pole. This volume contains nine chapters and an introduction.

I.4.1. *Chapters*

Chapter 1 proposes a Framework for Enterprise Interoperability, which aims at clarifying the enterprise interoperability concept, on the one hand, and defining and delimiting the enterprise interoperability domain, on the other hand. Three basic dimensions have been defined: Interoperability Concerns, which defines the content of interoperation that may take place at various levels of the enterprise; Interoperability Barriers, which identifies various obstacles to interoperability in three categories and Interoperability Approaches, which represents the different ways in which barriers can be removed.

Chapter 2 presents the various types of collaborations networked companies can sustain. Companies involved in a network can indeed choose various modes of exchange, depending on the level of involvement of the various partners. Advances in information management technology as well as the use of the Internet have altered the relationships of partners within a network, their nature evolving from strategic alliances to clusters of organizations and virtual communities based on professional social networks. Further into the chapter, we present a classification of the various types of collaboration in relation to the level or interoperability they require.

Chapter 3 focuses on the interoperability feature that has to be seen as a specific requirement in its own right. It highlights what seems to be relevant in terms of conceptual definitions and modeling framework whenever a (group) of engineer(s) intends to design what we call here a "natively interoperable system" or at least a system maximizing its abilities to be interoperable. To proceed, a clear definition of the concept of interoperability applied to complex systems is the first prerequisite. The second prerequisite consists in establishing the needs of a design team assigned to design such systems. An interface design pattern model with sufficient generic, formal and pragmatic qualities is proposed.

Chapter 4 aims to provide the adequate environment for the implementation of a method dedicated to the validation and verification of interoperability requirements. This method defines interoperability metrics as process Performance Indicators (PIs): the average elapsed time, average cost and percentage of failure at the business process level. In the first step of the approach proposed, a framework is used to specify some definitions of the method by making several choices in the way of applying the generic method. In the second step, the state of available data is compared with the data needed by the specific definition of the method, and thus the decision of implementing the latter can eventually be taken.

Chapter 5 presents the basic concepts and principles of decisional interoperability. This chapter does not take the socio-psychological aspect of decision-making into account and focuses on the formal and quantitative aspects in order to reduce the uncertainty, in terms of the objectives that the partners have to reach, during a collaborative decision-making process. Furthermore, it also aims at extending the GRAI decisional model concepts to the development of interoperability in the context of networked enterprises. On the basis of these works, a definition of decisional interoperability is proposed and some design principles to build a decisional interoperability solution are presented and described.

Chapter 6 first presents and discusses the research works within the interoperability measurement field. Then, a set of interoperability measurement components (potentiality, compatibility and performance measurement) are outlined and running examples are given to better understand metrics. Finally, three developed measurements and their possible improvement are discussed.

Chapter 7 aims to show how interoperability contributes to fulfilling the needs for transactional and/or informational collaboration in modern supply chains. After explaining the need for interoperability in supply chains, this chapter presents the various types of interoperability in a supply chain, as well as the main supply information systems and structures to support supply interoperability. The chapter ends with showing how much SaaS applications have disrupted the habits of supply chains and highlights the resulting benefits in terms of interoperability.

Chapter 8 discusses organizational interoperability issues between public and private organizations in an extended administration. Public administrations face the challenge to establish and manage networks to deliver public services, and this challenge questions the interoperability of public and private organizations that compose them. This chapter presents a framework, which can help manage the interoperability and evaluate the network performance, by identifying the specific interoperability requirements for a public–private network.

Chapter 9 focuses on the e-Health interoperability. The objective of this chapter is to make a review on initiatives promoting interoperability within the health sector in order to identify the limitations of existing work and contribute to fostering the interoperability of health systems. This study proposes a comparative analysis of the main frameworks for interoperability within the health domain with respect to the dimensions defined by the EIF and FEI. Some recommendations to deal with the identified limitations are proposed at the end of this chapter.

I.4.2. *Interpretative framework*

Three main types of contribution to the interoperability theme can be considered in this volume and in the forthcoming ones, which will complete the series.

The first type of contribution pertains to scientific investigation. There, one will find all one needs to define interoperability, to set the main principles from a general perspective or to delineate a specific sub-issue.

Chapters 1, 2, 5 and 6 pertain to that type of contribution.

The second type of contribution concerns technology and support. The chapters pertaining to that type of contribution tend to be less theoretical and propose directions of development, in a more downstream perspective.

In this context, we will find Chapters 3 and 4.

The third type of contribution focuses on applicative domains. It aims at presenting results applied to a specific economic or technical sector.

Chapters 7, 8 and 9 come under this category.

In this initial volume, we did not intend to balance the importance of these three types. Let us wager that a reasonable balance will gradually emerge following the publication of the forthcoming volumes of the series.

I.5. Bibliography

[CHA 15] CHAPURLAT V., DACLIN N., BILLAUD S., "An interface pattern model for supporting design of natively interoperable systems", *Enterprise Interoperability*, vol. 213, pp. 171–185, *6th International IFIP Working Conference, IWEI 2015*, Nîmes, France, 28–29 May 2015.

[CHE 03] CHEN D., DOUMEINGTS G., "European Initiatives to develop interoperability of enterprise applications – basic concepts, framework and roadmap", *Journal of Annual reviews in Control*, vol. 27, no. 2, pp. 151–160, 2003.

[IDE 03] IDEAS, IDEAS project deliverables (WP1-WP7), Public reports, available at: www.ideas-road map.net, 2003.

[IEE 90] IEEE, IEEE standard computer dictionary: a compilation of IEEE standard computer glossaries, 1990.

[VER 96] VERNADAT F.B., *Enterprise Modelling and Integration: Principles and Applications*, Chapman & Hall, London, 1996.

1

Framework for Enterprise Interoperability

1.1. Introduction

Enterprise interoperability means many things to many people. It is interpreted in many different ways in different contexts with different expectations [CHE 03, CHE 04]. This is not only true in industry but also in research communities and sometimes even within a working group. Without a clear and shared understanding on the precise meaning of interoperability, research and development efforts cannot be efficiently carried out and coordinated. The proposed Framework for Enterprise Interoperability aims, on the one hand, to clarify the enterprise interoperability concept, and, on the other hand, to define and delimit the enterprise interoperability domain [INT 07].

The Framework for Enterprise Interoperability was developed under European FP6 NoE INTEROP (Interoperability Research for Networked Enterprises Applications and Software) [INT 03]. It was inspired from or influenced by the following works: LISI (Levels of Information Systems Interoperability) [C4I 98], IDEAS Interoperability Framework [IDE 03], European Interoperability Framework [EIF 04] and ATHENA Interoperability Framework [ATH 03].

Enterprise interoperability defined in this framework is considered as the ability to (1) communicate and exchange information, (2) use the information exchanged and (3) access to functionality of a third system. This definition is based on some existing ones defined in [IEE 90, VER 96, IDE 03]. It has been considered that enterprise systems are not interoperable because of barriers to interoperability. Barriers are incompatibilities of various kinds at the various enterprise levels. There exist common barriers to interoperability and generic solutions to remove those barriers.

Chapter written by David CHEN.
This chapter is based on the paper [CHE 09].

1.2. Enterprise interoperability concepts

There are many concepts relating to interoperability. This section focuses on the basic concepts for the purpose of helping to understand the framework for enterprise interoperability and its underlying domain. These concepts are identified based on the state-of-the-art of relevant interoperability researches [INT 04].

1.2.1. *Interoperability barriers*

Interoperability barriers are a fundamental concept for understanding interoperability problems. Many interoperability problems are specific to particular application domains. These can be issues like support for particular attributes or particular access control regimes. However, general barriers or problems of interoperability exist and can be identified; and most of them being already addressed [KAS 04, EIF 04, ERI 04]. Consequently, the objective is to identify and categorize those common barriers to interoperability. The term "barrier" means an "incompatibility" or "mismatch" that obstructs the sharing and exchanging of information. Three categories of barriers are identified: (a) *conceptual*, (b) *technological* and (c) *organizational*.

a) *Conceptual barriers*

Conceptual barriers are concerned with the syntactic and semantic incompatibilities of information to be exchanged. These problems concern the representation of information at the high level of abstraction (such as, for example, the enterprise models of a company) as well as the level of programming (for example, low capacity of semantic representation of XML):

– syntactic incompatibility can be found whenever different people or systems use different structures (or format) to represent information and knowledge. To tackle this problem, the UEML initiative [UEM 01] developed a neutral model to allow mapping between different enterprise models built with different syntaxes;

– semantic incompatibility is seen as an important barrier to interoperability as the information and knowledge represented in most of the models or software have no clearly defined semantics to allow unambiguous understanding of the meaning of information. At the current stage of research, the most known technique to solve this problem is semantic annotation and reconciliation using ontology.

Conceptual barriers are the main barriers to interoperability.

b) *Technological barriers*

Technological barriers are concerned with the use of computers or ICT (Information and Communication Technology) to communicate and exchange information. Typical technological barriers are for example incompatibility of IT architecture & platforms, infrastructure, operating system, etc. From a purely technical perspective, these problems concern the ways to present, store, exchange, process and communicate data and information through the use of software systems. Examples of technological barriers are:

– communication barriers, e.g. incompatibility of the protocols used to exchange information;

– content barriers, e.g. different techniques and methods used to represent information, or incompatibility in the tools used to encode/decode the information being exchanged;

– infrastructure barriers, e.g. use of different incompatible middleware platforms.

Technological barriers are additional barriers with respect to conceptual ones. They are caused by the use of computer. Technological barriers only exist if computers are involved in an interoperation.

c) *Organizational barriers*

Organizational barriers are concerned with the incompatibilities of organization structure and management techniques implemented in two enterprises. For example, one organization structure barrier is related to the way of assigning responsibility and authority. Indeed, if two companies have different organization structures (e.g. hierarchical vs. networked) and management techniques, some necessary mappings may need to be done before the two sides become interoperable at an operational level:

– responsibility needs to be defined to allow two parties knowing who is responsible for what (process, data, software, etc.). If responsibility in an enterprise is not clearly and explicitly defined, interoperation between two systems may be obstructed;

– authority is an organizational concept which defines who is authorized to do what. For example, it is necessary to define who is authorized to create, modify, maintain data, processes, services, etc.;

– organization structure refers to the style by which responsibility, authority and decision-making are organized. For example, we can talk about centralized versus decentralized organizations, or hierarchical versus matrix or networked organization structures.

Organizational barriers are also additional barriers. Compared with conceptual barriers (centered on information problems) and technological barriers (concerned with IT problems), organizational barriers originate from the problems of humans.

1.2.2. *Interoperability concerns*

This section defines the interoperations that can take place from the various concerns (or viewpoints) of the enterprise. Although the definitions are mainly given from the point of view of IT-based applications, they apply to non-computerized systems as well. This categorization is based on the ATHENA Technical Framework [GUG 05].

The interoperability of communication is a basic condition to allow for interoperability to take place. It relates primarily to the interconnection of systems and equipment as well as communication means. From an IT point of view, it is concerned with communication protocols (i.e. from cable connection to the protocol of layers 1 to 4 of the OSI model) and interfaces (layers 5 to 7 of the OSI model). Since the topic of interoperability of communications is already widely discussed, it will not be a subject of discussion in this section.

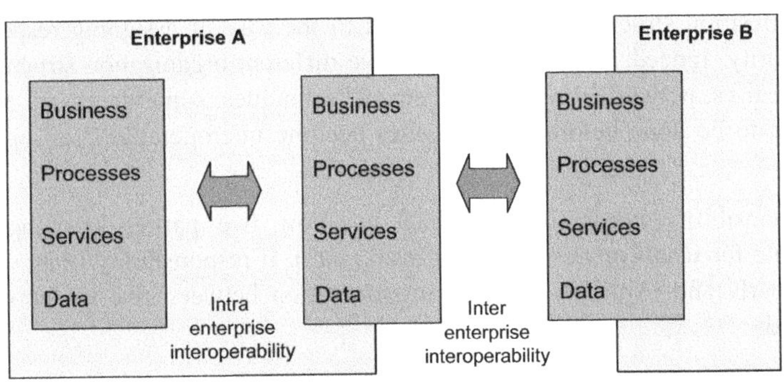

Figure 1.1. *Adapted from [ATH 07]*

In the domain of enterprise interoperability, the following four interoperability concerns are identified as shown Figure 1.1: (a) data, (b) service, (c) process and (d) business. In an enterprise, data is used by services (or functions to provide a service). Services (functions/activities) are employed by processes to realize business of the enterprise. From another point of view, the goal of an enterprise is to run its business. To realize the business, we need processes. Processes employ services/functions that in turn need data to perform activities.

a) *Interoperability of data*

The interoperability of data refers to the simultaneous operation of different data models (hierarchical, relational, etc.) and the use of the different query languages. Moreover, their content is organized according to conceptual schemas (i.e. vocabularies and sets of structures of data) that are related to particular applications. The interoperability of data is concerned with finding and sharing information coming from heterogeneous databases, and which can, moreover, reside on different machines with different operating systems and database management systems.

Data interoperability plays an essential role in enterprise interoperability. It is concerned with the ability to exchange both non-electronic data (documents) and machine transportable data (data files, data stored in a database) and use the data/information exchanged. Data interoperation may occur when two partners simply exchange two data files (e.g. Excel files); or in the case of process interoperability or service interoperability themselves employ data to perform activities. Typical barriers which prevent data interoperations are for example conceptual ones, such as different semantics and syntax to represent data, and also technological one (different database technologies and coding techniques) and organizational ones (database management, security policy, etc.).

b) *Interoperability of service*

The interoperability of service is concerned with identifying, composing and operating various applications (designed and implemented independently) by solving the syntactic and semantic differences as well as by finding the connections to the various heterogeneous databases. The term “service” is not limited to IT applications, but includes functions of the company or of networked enterprises.

Service interoperability deals with the capability of exchanging services (works) among partners. Service interoperability has two main problems: service exchange between a service demander and a service provider; interconnection between different services to form a complex service (the last case is related to process interoperability as well). A service is provided by a resource (computer type, machining type and human type). Issues relating to service interoperability are concerned with the description (both from the syntax and semantic aspects) of the services required and provided, the mechanisms to search and discover a distributed service provider, the ICT supports for service discovery, composition and the organizational issues relating the management of service exchange, etc.

c) *Interoperability of process*

The interoperability of processes aims at making different processes work together: a process defines in which order services (functions) are related. Generally in a company, several processes run in interactions (in series or parallel). In the case of the networked enterprise, it is also necessary to connect the internal processes of two companies to create a common process.

More precisely, process interoperability arises when linking different process descriptions (be they documents or supported by software) to form collaborative processes and perform verification, simulation and execution. Usually, different process description languages are used to define different process models for different purposes. Typical barriers that prevent process interoperations include: different semantics and syntax used in different process modeling languages; incompatible process execution engines and platforms; different process organization mechanisms, configurations and managements. Developing process interoperability means to find solutions to allow mapping, connecting, merging, and translations of heterogeneous process models, applications and tools.

For interoperability reasons, these solutions are concerned with the connection points of the processes and not with the processes as a whole. The latter would lead to process integration.

d) *Interoperability of business*

The interoperability of business refers to work in a harmonized way at the levels of organization and company in spite of for example, the different modes of decision-making, methods of work, legislations, culture of the company and commercial approaches, etc., so that business can be developed and shared between companies.

In other words, business interoperability is concerned with how business is understood and shared without ambiguity among interoperation partners. Business interoperability explores interoperability from a business perspective and identifies the fundamental artifacts related to business issues. These issues range from the business vision and culture to the ICT infrastructure support as well as the compatibility between different organization structures, methods of work, accounting systems and rules, labor legislations, etc. Developing business interoperability is a means to find ways to make those issues be harmonized or at least understood through necessary mappings and negotiations. It is worth noting that for interoperability reasons, these solutions are concerned with the connection points of the business not with the business as a whole. The latter would lead to business integration.

1.2.3. *Interoperability approaches*

Defining the enterprise interoperability domain is not only a matter of identifying barriers and solutions for removing barriers, but also the ways in which these barriers are removed. Establishing interoperability requires relating entities in some way. According to ISO 14258 (concepts and rules for enterprise models) [ISO 99], there are three basic ways to relate entities together: (a) integrated, (b) unified and (c) federated, as shown in Figure 1.2.

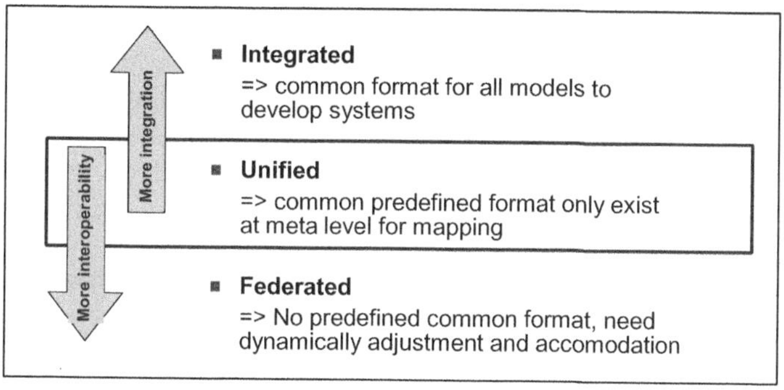

Figure 1.2. *Basic approaches to develop interoperability*

a) *Integrated approach*

Developing interoperability through an *integrated approach* means that there exists a common format for all models. Diverse models are built and interpreted using/against the common template. This format must be as detailed as the models themselves. The common format is not necessarily an international standard but must be agreed by all parties to elaborate models and build systems. Example of developing interoperability using an integrated approach is ebXML.

The integrated approach is more relevant to full integration rather than full interoperability. This approach is suitable when designing and implementing new systems rather than re-engineering existing systems for interoperability. To some extent, the integrated approach is more adapted to developing intra-enterprise interoperability rather than inter-enterprise.

The integrated approach ensures the global consistency and coherence of the system. Various components of the system are designed and implemented using a common format (or standard) so that interoperability is seen as designed-in quality (of the system's components). Interoperation between various parts can be obtained *a priori* without any interfacing effort.

b) *Unified approach*

Interoperability can also be established using a *unified approach*. It means there is a common format but it only exists at the meta-level. This format is not an executable entity, as is the case in an integrated approach. Instead, it provides a mean for equivalence to allow mapping between models and applications. For example, using the meta-model, a mapping between the constituent models is possible even though they might encounter loss of some semantics or information.

Most research results developed in the domain of interoperability have adopted the unified approach. For example, UEML (Unified Enterprise Modeling Language) aims at defining a neutral format at meta-modeling level to allow mapping between enterprise models and tools. The STEP initiative elaborated in ISO TC184 SC4 also defined a neutral product data format at the meta-modeling level to allow various product data models exchanging product information.

The unified approach is particularly suitable for developing interoperability for collaborative or networked enterprises. To be interoperable with networked partners, a new company just needs to map its own model/system to the neutral meta-format without the necessity to make changes on its own model/system. This approach presents the advantage of reduced efforts, time and cost in implementation. It is also adapted to the situation where a large company needs to interoperate with SMEs. Normally, a SME works with more than one big company; to interoperate with different companies, the unified approach seems to be a suitable solution.

c) *Federated approach*

In the case of using a *federated approach*, there is no format which is common to all. To establish interoperability, parties must accommodate and adjust "on the fly". Using the federated approach implies that no partner imposes their models, languages and methods of work. They must negotiate dynamically to reach the needed agreements. This means that they must share some ontology.

The federated approach can also make use of meta-models for mapping between diverse models/systems. The difference to a unified approach is that this meta-model is not pre-defined but established "dynamically" through negotiation. Consequently, this approach is more suitable for "peer-to-peer" situations rather than the cases mentioned in the unified approach. It is particularly adapted to Virtual Enterprises, where diverse companies join their resources and knowledge to manufacture a product with a limited duration.

Using the federated approach to develop enterprise interoperability is the greatest challenge and little activity has been performed in this direction. A main research area is the development of a "mapping factory" which can generate on-demand customized AAA (Anybody-Anywhere-Anytime) mapping agents among existing systems. It is worth noting that a specific support for the federated approach is seen in entity profiles which identify particular entity characteristics and properties relevant for interoperation (ISO 15745 and ISO 16100).

All three approaches allow for developing interoperability between enterprise systems. The federated approach is considered as the most interesting to achieve full interoperability. However, the choice depends on the context and requirements. If the need for interoperability comes from a merger of enterprises, the integrated approach seems to be the most adapted one. In this case, there is only one common format for all systems, and all models are built and interpreted according to this one. If the need for interoperability concerns a long-term collaboration, the unified approach would be a possible solution. For that, a common meta-model across partners' models provides a means for establishing equivalence allowing for mapping between diverse models. Finally, if the need for interoperability originated from a short-term collaboration project (e.g. virtual enterprise), the federated approach can be used. Partners must dynamically adapt to find necessary agreements.

Figure 1.3 summarizes enterprise interoperability concepts with the aim to define the ontology of enterprise interoperability (adapted from [NAU 07]).

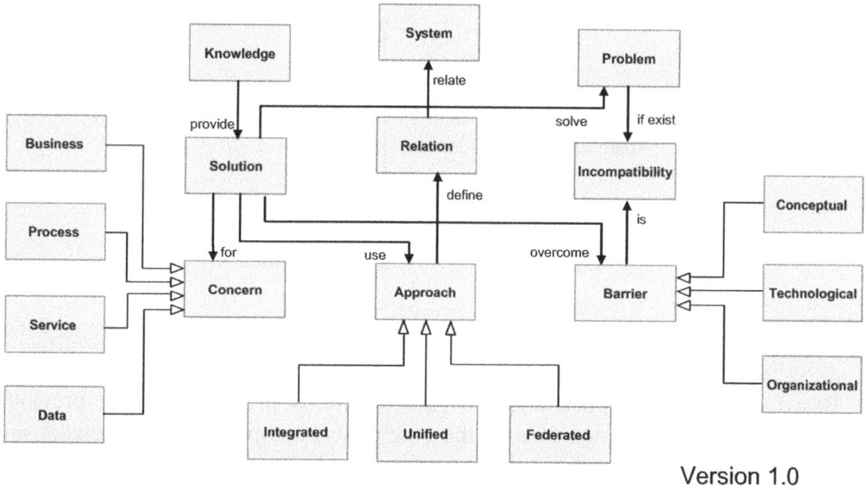

Figure 1.3. *Towards an ontology of enterprise interoperability*

1.3. Framework for Enterprise Interoperability

The term "framework" is a conceptual model. It refers to an organizing mechanism for structuring and categorizing "things" relative to a domain. A framework does not provide an operational solution to solve a business problem. The interoperability framework presented in this chapter aims at structuring the main concepts of the enterprise interoperability domain. The framework has three basic dimensions: interoperability concerns, interoperability barriers and interoperability approaches.

1.3.1. *Problem space versus solution space*

The first two dimensions: interoperability concerns and interoperability barriers constitute the problem space of enterprise interoperability (see Figure 1.4). The intersection of an interoperability barrier and an interoperability concern is the set of interoperability problems having the same barrier and concern. The three dimensions together constitute the solution space of enterprise interoperability. The intersection of an interoperability barrier, an interoperability concern and an interoperability approach is the set of solutions to breakdown a same interoperability barrier for a same concern and using a same approach.

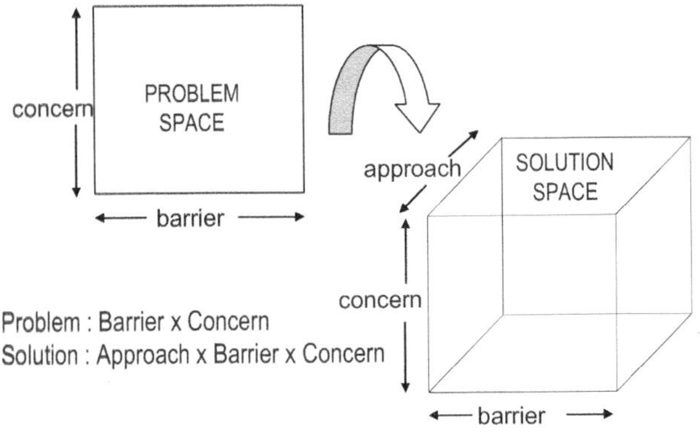

Figure 1.4. *Problem versus solution spaces*

1.3.2. *The two basic dimensions*

Based on these concepts, and the categorizations presented in the previous section, the two basic dimensions of the enterprise interoperability framework are shown in Figure 1.5: (1) interoperability concerns and (2) interoperability barriers.

Some examples of interoperability barriers and problems are also shown in the figure. For example, at the intersection of conceptual barrier and process concern, one of the interoperability problems encountered is for example the impossibility of exchanging process model information between IDEF3 and BPMN models because of the syntax incompatibility of the two models.

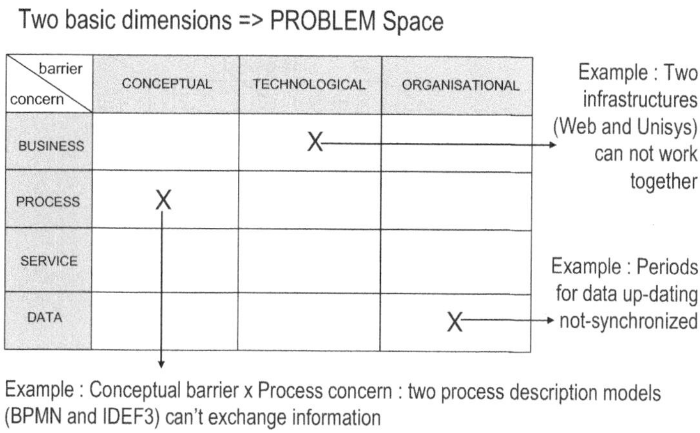

Figure 1.5. *Framework for Enterprise Interoperability (two basic dimensions)*

Adopting the barrier-driven approach to tackle interoperability problems implies that research is bottom-up to find knowledge and solutions to remove barriers. Although the three categories of barriers concern all the four levels, conceptual and organizational barriers are more important at the higher levels while technological (barriers due to the use of ICT) ones have a greater impact on the lower levels.

1.3.3. *The third dimension*

The third basic dimension (interoperability approaches) added to the two dimensional framework allows for categorizing solutions of enterprise interoperability according to the ways of removing the barriers. As discussed in the previous section, this dimension considers the three basic approaches to develop interoperability: integrated, unified and federated.

The framework, with its three basic dimensions, is shown in Figure 1.6. We recall the three basic dimensions as follows:

– *interoperability concerns* that define the content of interoperation that may take place at various levels of the enterprise (data, service, process, business);

– *interoperability barriers* that identify various obstacles to interoperability in three categories (conceptual, technological and organizational);

– *interoperability approaches* that represent the different ways in which barriers can be removed (integrated, unified, and federated).

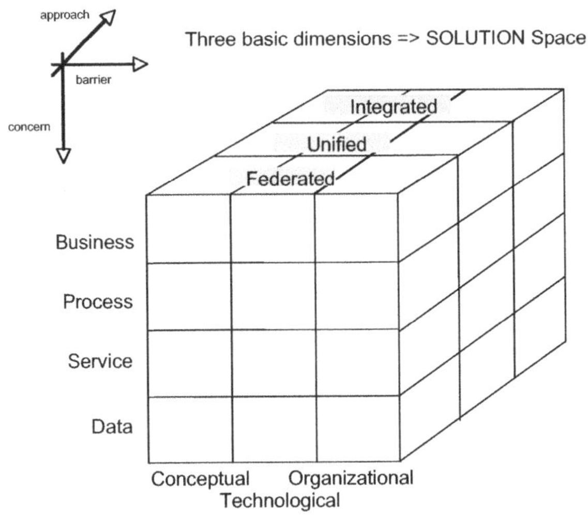

Figure 1.6. *Framework for Enterprise Interoperability*

The third dimension allows for capturing and structuring interoperability solutions according to their ability to remove interoperability barriers. For example, Process Specification Language (PSL) [SCH 00] is a solution that contributes to removing the *conceptual* barrier (both syntax and semantics) concerning *process* through a *unified* approach. Figure 1.7 shows the position of a PSL solution in the framework.

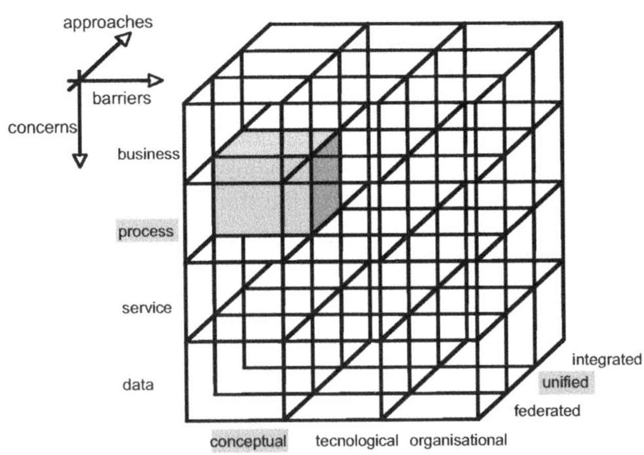

Figure 1.7. *Position of a PSL in the framework*

To help capture the solutions and their positioning in the framework, a template is proposed. Figure 1.8 shows a simplified example of the template to describe a PSL solution.

Interoperability knowledge/solution template	
Name of the knowledge/solution: PSL	
1. Interoperability concern	Process level
2. Interoperability barrier	Conceptual (Syntax and sematics)
3. Interoperability approach	Unified approach
4. Interoperability problem	different process models use different process languges and are not interoperable
5. Interoperability knowledge	Define a neutral Process Specification Language (PSL) and related ontology as a metamodel to allow mapping between different process models
6. Example (optional)	
7. Remarks	Initially proposed by NIST, now moved to standardisation at ISO level
8. References	ISO CD 18629 (2001), Industrial automation systems and integration, Process Specification Language (PSL), JW8/ISO 184/SC4/SC5

Figure 1.8. *Template example to collect a PSL solution*

1.3.4. *Complementary dimensions*

The third dimension (interoperability approaches) discussed above can be replaced by other complementary dimensions according to the objective related to a particular usage of the framework. In other words, the third dimension can be considered as an open dimension. At the current stage of this research, there are three possible dimensions to be identified.

Interoperability engineering dimension

This dimension aims at defining a set of phases (steps) to follow for establishing interoperability between two enterprises (or any two business entities). The system life cycle phases defined in ISO 15704 Standard [ISO 00] have been adapted and three main phases have been defined as follows: (1) requirements definition, (2) design specification and (3) implementation. Figure 1.9 shows this dimension in relation to the two basic dimensions of the framework.

Using this framework in an interoperability project, the requirements definition phase identifies possible barriers to interoperability that exists between two enterprises and the interoperability concerns to be addressed. At the design specification phase, search/develop interoperability solutions for removing the barriers is defined. The implementation phase allows for implementing and testing the solutions and measuring the interoperability achieved.

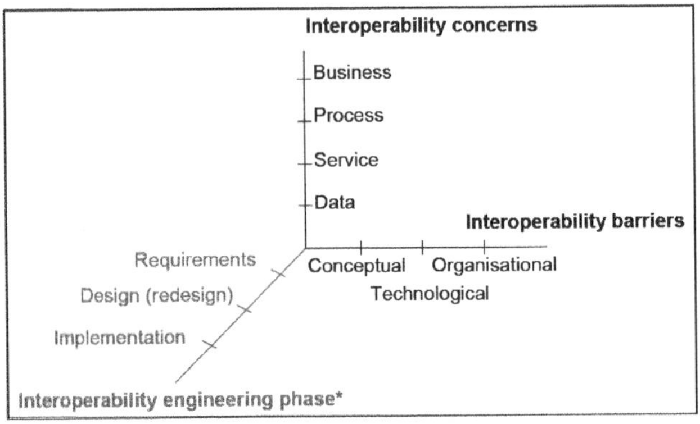

Figure 1.9. *Interoperability engineering phase dimension*

Interoperability measurement dimension

The degree of interoperability is a measure characterizing the ability of interoperation between two enterprises (or systems). At the current stage of research, three types of measurement are identified as shown in Figure 1.10: (1) interoperability potentiality measurement, (2) interoperability compatibility measurement and (3) interoperability performance measurement. The interoperability degree of a given enterprise (or a system) can be defined by a vector characterized by the three measurements [DAC 06].

The potentiality measurement is concerned with the identification of a set of system properties that have impact on the interoperability. This measure is performed on one enterprise/system without knowing its interoperation partner. The objective is to evaluate the potentiality of a system to adapt accommodate it to overcome various possible barriers. For example, an open system has a higher potential of interoperability than a closed system.

The compatibility measurement is performed when the partner/system of the interoperation is known. The measure is done with respect to the identified barriers to interoperability. The highest degree means there is no barrier to interoperability. The inverse situation means the poorest degree of interoperability.

The performance measurement is performed at run time to evaluate interoperations between two cooperating enterprises (or systems). Criteria such as cost, delay and quality can be used to measure the interoperation performance. Therefore, each type of measurement has to be valued with local coefficients in order to get a global coefficient ranging from "poor interoperability" to "good interoperability".

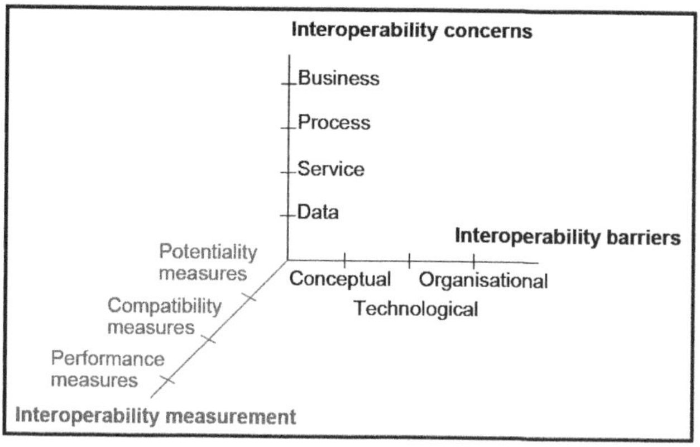

Figure 1.10. *Interoperability measurement dimension*

Interoperability knowledge dimension

From an abstraction level point of view, there are two types of knowledge/solution: conceptual or technological, as shown in Figure 1.11. The notion of a conceptual versus technological solution also comes from engineering design where we distinguish between conceptual design and technical design. Conceptual design aims at specifying a conceptual solution that is independent from technology for implementing the solution. For a given conceptual solution, there may exist several different technologies to implement the solution. The technology choice is made at technical design stage. This technical design is also called the detail design.

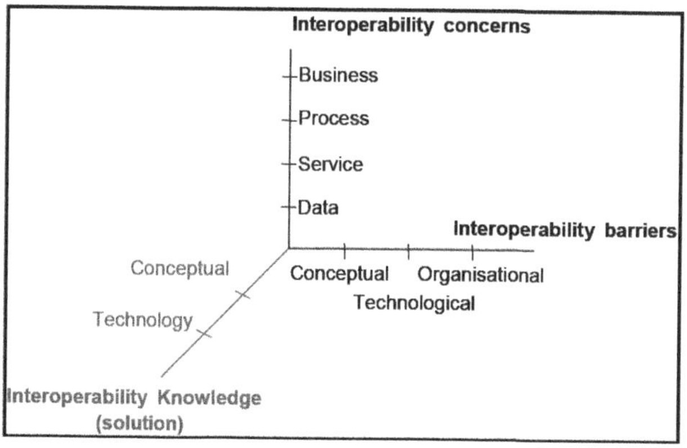

Figure 1.11. *Interoperability knowledge/solution dimension*

Conceptual solution also means the conceptual description of a solution. In this case, a conceptual solution describes the "ideas" that allow solving a problem without specifying how to concretize the "ideas", i.e. how to implement the "ideas".

Conceptual solution can also be a conceptual representation of an existing technical solution. In this case, only generic aspects of the solution (for example functions) are "filtered" and represented without specific technological details.

With this dimension, it is possible to position interoperability knowledge (solution) in the framework in a more precise way. For each category of barriers (conceptual, technological and organizational), solutions can be conceptual, technological or both. For example, the "semantic annotation" is a method to move semantic barrier concerning the all four levels (conceptual knowledge/conceptual barrier/all concerns). A* tool developed in ATHENA project is a technological solution to remove semantic barrier concerning all the concerns (technology knowledge/conceptual barrier/all concerns).

1.4. Conclusion and prospects

This chapter presented a Framework for Enterprise Interoperability in which the problem and solution spaces defined the domain of enterprise interoperability. With respect to the existing interoperability frameworks (ATHENA, IDEAS, EIF, etc.), the proposed framework is barrier driven. If a problem or a solution cannot find its place in the framework, it is not an enterprise interoperability problem or solution. The complementary dimensions allow for "personalizing" the framework according to a specific user's need.

Establishing interoperability means relating two (or more) systems and removing any incompatibilities. Incompatibility is the fundamental concept used in defining the scope of interoperability domain. The concept "incompatibility" has a broad sense and is not only limited to "technical" aspects as usually considered in software engineering, but also "information" and "organization", and concerns all levels of an enterprise. Another fundamental consideration is the generic characteristic of the interoperability research. Indeed, there are generic problems and solutions regardless of the content of information exchanged between two systems.

Benefits and applications of the framework include a better understanding of enterprise interoperability research problems. Furthermore, knowledge and solutions can be structured in the framework to allow gap analysis, so that future research orientation can be defined to close the gaps. It also contributed to the standardization work in this area. The International Standard EN/ISO 11354 (Framework for

Enterprise Interoperability) elaborated jointly by CEN TC310/WG1 and ISO TC184 SC5/WG1 has adopted the proposed framework.

1.5. Bibliography

[ATH 03] ATHENA, Advanced Technologies for Interoperability of Heterogeneous Enterprise Networks and their Applications, FP6-2002-IST-1, ATHENA Integrated Project, Technical Annex, 2003.

[ATH 07] ATHENA, D.A4.2: Specification of Interoperability Framework and Profiles, Guidelines and Best Practices, ATHENA Integrated Project, Deliverable D.A4.2, March 2007.

[C4I 98] C4ISR, Architecture Working Group (AWG), Levels of Information Systems Interoperability (LISI), 1998.

[CHE 03] CHEN D., DOUMEINGTS G., "European Initiatives to develop interoperability of enterprise applications – basic concepts, framework and roadmap", *Journal of Annual reviews in Control*, vol. 27, no. 2, pp. 151–160, 2003.

[CHE 04] CHEN D., VERNADAT F., "Standards on enterprise integration and engineering – a state of the art", *International Journal of Computer Integrated Manufacturing*, vol. 17, no. 3, pp. 235–253, 2004.

[CHE 09] CHEN D., "Framework for enterprise interoperability", *Proceedings of the Congrès International de Génie Industriel (CIGI2009)*, Bagnères de Bigorre, France, 10–12 June 2009.

[DAC 06] DACLIN D., CHEN D., VALLESPIR B., "Enterprise interoperability measurement – basic concepts", *Workshop EMOI'06 in Conjunction with CAiSE'06*, Luxemburg, 2006.

[EIF 04] EIF, European Interoperability Framework, White Paper, available at: http://www.comptia.org, 2004.

[ERI 04] ERISA, A guide to Interoperability for Regional Initiatives, The European Regional Information Society Association, Brussels, 2004.

[GUG 05] GUGLIELMINA C., BERRE A., Project A4 (Slide presentation), ATHENA Intermediate Audit at 29–30 September 2005, Athens, Greece, 2005.

[IEE 90] IEEE, IEEE standard computer dictionary: a compilation of IEEE standard computer glossaries, 1990.

[IDE 03] IDEAS, "IDEAS project deliverables (WP1-WP7)", Public reports, available at: www.ideas-road map.net, 2003.

[INT 03] INTEROP, Interoperability Research for Networked Enterprises Applications and Software, Network of Excellence, Project Proposal Part B, 2003.

[INT 04] INTEROP, Knowledge map of research in interoperability, Deliverable D1.1, INTEROP NoE, WP1, version 1, 2004.

[INT 07] INTEROP, Enterprise Interoperability-Framework and knowledge corpus – final report, Research report of INTEROP NoE, FP6 – Network of Excellence – Contract no. 508011, Deliverable DI.3, 2007.

[ISO 99] ISO 14258, Industrial Automation Systems – Concepts and Rules for Enterprise Models, ISO TC184/SC5/WG1, 1999-April-14 version, 1999.

[ISO 00] ISO 15704, Industrial automation systems – Requirements for enterprise-reference architectures and methodologies, 2000.

[ISO 04] ISO 19440, Enterprise integration – Constructs for enterprise modelling (CEN/ISO 19440), CEN TC 310-ISO/TC 184, 2004.

[KAS 04] KASUNIC M., ANDERSON W., Measuring systems interoperability: challenges and opportunities, Software engineering measurement and analysis initiative, Technical note CMU/SEI-2004-TN-003, 2004.

[NAU 07] NAUDET Y., Integrating the Enterprise Interoperability Framework into the Ontology of Interoperability, Research Note, 2007.

[SCH 00] SCHELENOFF G., GRUNINGER M., TISSOT F. *et al.*, The Process Specification Language (PSL): Overview and Version 1.0 Specification, National Institute of Standards and Technology (NIST), Gaithersburg, MD, 2000.

[UEM 01] UEML, IST – 2001 – 34229, Unified Enterprise Modelling Language (UEML), Thematic Network, Annex 1, Description of Works, 2001.

[VER 96] VERNADAT F.B., *Enterprise Modelling and Integration: Principles and Applications*, Chapman & Hall, London, 1996.

2

Networked Companies and a Typology of Collaborations

2.1. Introduction

The notion of interoperability is only relevant when there is some degree of networking or collaboration between companies. This statement may seem obvious, but it underlines the fact that the issue of enterprise interoperability emerged and developed with the growth of economic networks of planetary scale. Speaking of networks is, however, insufficient because there exist many ways to set up exchanges and collaborations between companies. These many ways lead to networks of various types and thus to various needs of interoperability.

This chapter aims to present the main types of network in order to infer, in fine, the type of interoperability needed in each particular case.

2.2. Various types of collaboration between companies

The forms of collaboration between networked companies may be varied, but they all at least involve organization of the partners' activities in a manner which specifies the roles of each participant and the rules by which to run the network. This type of organization is called Collaborative Network Organization (CNO) [CAM 08a] and can take several forms:

– strategic alliance;

– integrated logistics management;

Chapter written by Séverine BLANC SERRIER, Yves DUCQ and Bruno VALLESPIR.

– network enterprise;

– virtual organizations and clusters.

2.2.1. *Strategic alliances*

Strategic alliance is a form of inter-enterprise coordination which rests midway between full internalization (hierarchical coordination) and full externalization (market coordination) [INS 01]. Strategic alliances constitute a specific case of an inter-organization relationship which aims to "do together" rather than simply "doing" (internalization) or "making do" (externalization).

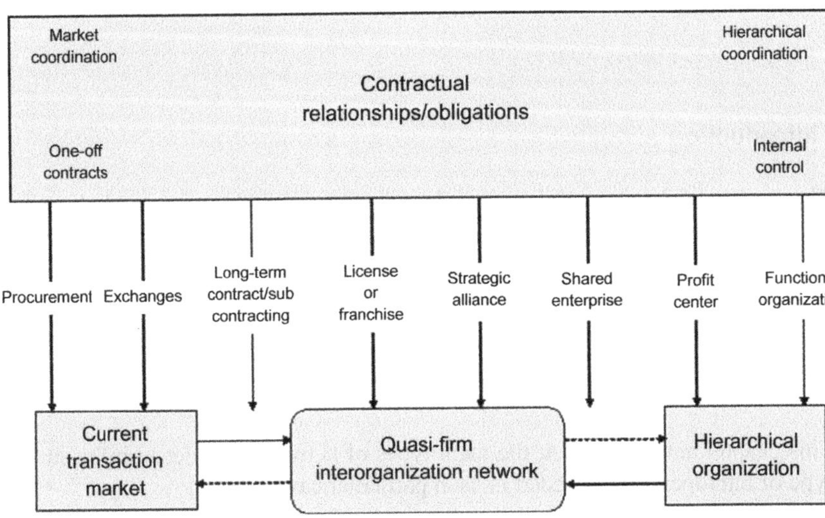

Figure 2.1. *Various forms of inter-organization relationships [JUL 94]*

There are many forms of inter-organization relationships, which span the whole continuum between the market and the hierarchy. Figure 2.1 shows these various forms in a continuum within which it positions strategic alliance [JUL 94].

For an alliance to qualify as strategic, it must involve at least one key resource of the company. It can, for example, involve its technology, business brands, patents, and means of production or its distribution networks. If we summarize the various aspects and dimensions of an inter-organization relationship, we can thus define a strategic alliance as: "A long term explicit agreement thanks to which independent companies exchange, share or combine strategic resources to achieve one or several goals, in order to build or consolidate advantages or favorable competitive positions" [INS 01]. This definition underlines the formal aspect of the agreement, its long-term reach, the independence of the partners, the involvement of strategic resources (which can be human, financial, technical, etc.) as well as the improvement of the competitive positions of each involved company. Alliances can have characteristics including several other aspects. They can be bilateral (two partners) or multilateral (several partners), symmetric or asymmetric (relatively to the respective size and importance of each partner), structured or not, anti-competitive or proto-competitive and, finally, horizontal or vertical [INS 01].

2.2.2. *Integrated logistics management*

Whereas marketing makes products adapt to the market, logistics and supply chains make "service" to the market adapt by allowing offering, at the lowest cost possible, a quality product at the precise location and time where the demand exists [INS 01]. Figure 2.2 outlines the logistics procedure consisting, first and foremost, of managing, through operational and transactional activities, the physical flows between the three classical subsystems presented by [CHE 93]:

1) supply logistics (upstream);

2) production logistics (internal);

3) distribution logistics (downstream).

Each subsystem is also composed of a range of logistical activities which can be of the operational or transactional type. In parallel with these physical flows, information flows ensure the running and coordination of the physical flow chain. The value chain (which does not appear in the diagram) encompasses this space of organization and coordination of logistic flows and describes the economic flows of value creation from the raw material to the end user. The value chain enables us to understand where, along the logistic chain (supply chain), value is created. At this level, we find the planning and development (or innovation) activities.

Integrated supply chains, as outlined in Figure 2.2, therefore involve three different levels of integration [CHE 93]. At the first level, integrated supply chains coordinate the various logistic subsystems, namely upstream logistics (supply), internal logistics (production) and downstream logistics (distribution). The various activities of these logistical sub-systems, such as stock management (upstream), storage of products in process (of production) and the delivery of end products to the customer (downstream) are thus organized in a broader coherent system, the integrated supply chain function.

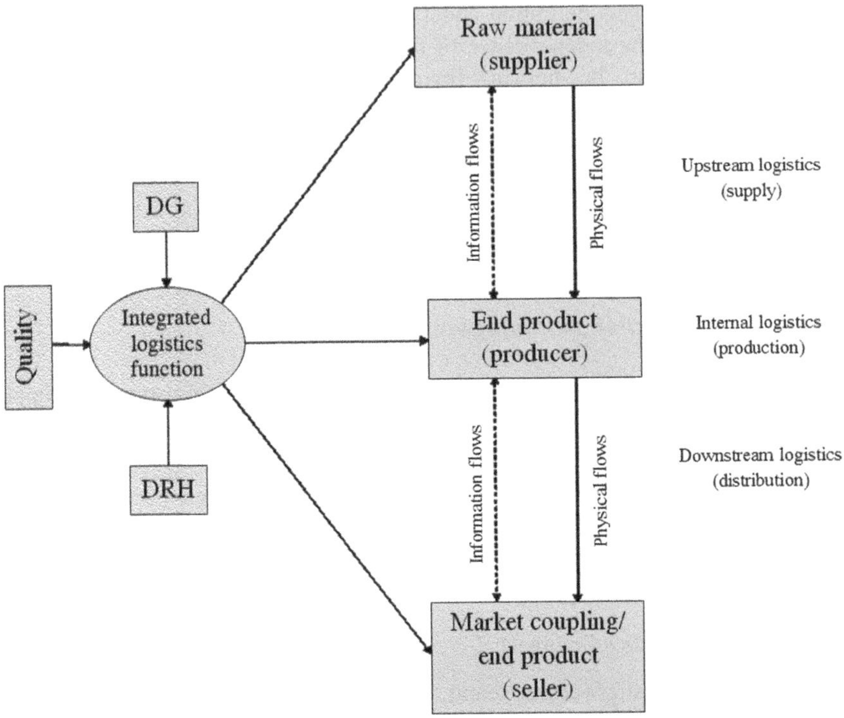

Figure 2.2. *Integrated supply chain [CHE 93]*

At the second level, logistics are integrated interfacing with the other functions of the company. Logistics, for example, participates with production in determining the location of products and, with the research and development department, in the design of packaging [COL 96]. Eventually, at the third level, integrated logistics contributes to inter-enterprise coordination from upstream to downstream. From the various suppliers (upstream) through to the various customers (downstream) via internal production, logistics must integrate all the physical and informational flows within what is conventionally called the supply chain. This external interdependency refers to the concept of supply chain management defined as an integrative approach aiming at the planning and monitoring of flows of products from suppliers to end customers. We will detail this aspect further on in this book.

Focusing on flows of products and on the inter-organization relationship, supply chain management prioritizes improvement of end-customer satisfaction over the meeting of specific marketing goals. Supply chain management, like total quality management, is truly a management philosophy. Since it takes all the agents into account, this more cooperation-oriented approach aims to reduce the opportunistic behaviors of each agent and to globally favor the satisfaction of the end customers [INS 01].

2.2.3. *Network enterprise*

Reacting to perturbations that lead purchasers to modify their relations with their suppliers, some companies went beyond adjusting their processes and tried to invent new forms of organizing which would be more flexible, more reactive, indicate better performance and may even be more normal [ESN 96]. To describe these new forms of organization, [ESN 96] uses the network metaphor, when others may speak of the virtual enterprise or of the platform organization. According to [ESN 96], since the economy of flexibility was discovered, the large, pyramidal and monolithic companies have lost terrain to more "federative" structures within which smaller organizations partner up in various ways, both vertically (along the value chain) and horizontally (to expand their international markets).

A network enterprise (or collaborative enterprise) is a network of legally independent companies linked to each other by the production cycle (of one or several products or a range of products). They are thus mutually dependent on a network of selected partners that evolves around a central core or a pivot company [JUL 94]. Despite this central core, we speak of "de-hierarchization" in the sense that the purchaser/supplier relationship shifts from a traditionally hierarchical relationship to collaboration.

This intermediary mode of organization (between the hierarchy and the market) is based on mutual trust and has the advantage of a systematic sharing of information that generates innovation by means of a creative synergy phenomenon (Figure 2.3). Not all companies of the network have the same importance. The "pivot company", which is usually a large manufacturer, does not collaborate in the same way with all the companies of the network. At least three different levels can be identified. We then speak of the shift from *capacity subcontracting* (linked to the market) to *specialty subcontracting* (hierarchical) and eventually to *intelligence subcontracting* (network): "First of all, one finds capacity subcontracting whereby the subcontractors meticulously complies to the specifications given by the contractor, […] then there is the specialty subcontractor who participates, to some degree, to the definition of the specifications of the ordered product […] and at a third level one finds the subcontractor of 'intelligence' who systematically attempts to design and develop the product best suited to meet the needs expressed by the contractor" [JAC 94].

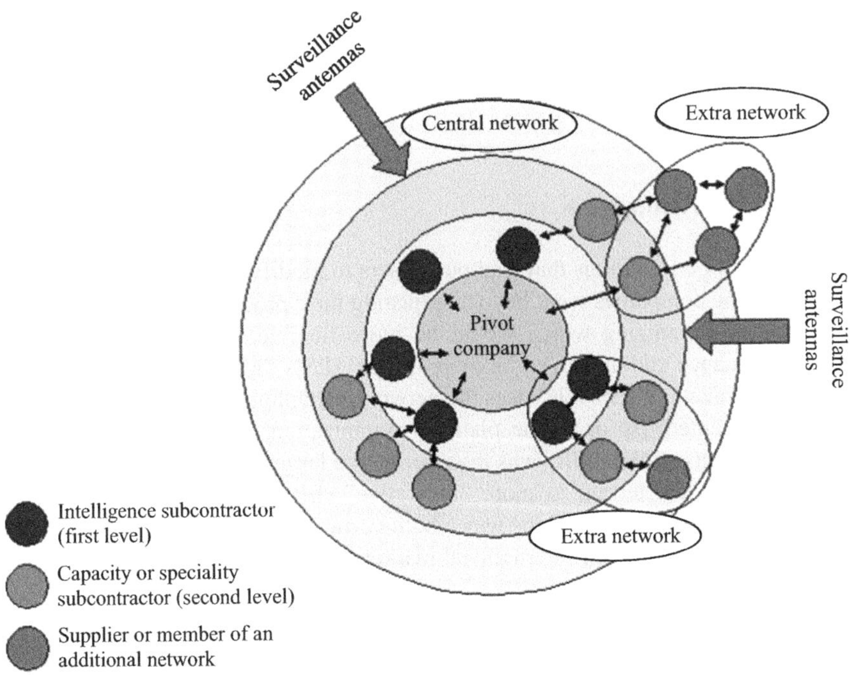

Figure 2.3. *The network enterprise*

These three levels of subcontracting are defined relative to the central network: centered on the pivot company and its various products or product ranges. However, as Figure 2.3 shows, various additional networks can be attached to the central network. The leadership of these additional networks then becomes the responsibility of the most advanced company, which is the closest to the core. Figure 2.3 presents a synthetic sketch of the network enterprise [JUL 94]. As an enterprise, to be part of a network does not imply being closed off the market. Several partners within the network can collaborate in that organization while remaining competitors on the market. This sometimes induces reluctance to partake in diffuse innovation. It is important to strike a balance between competition and collaboration in order to promote innovation through the need to do better (due to competition) and to boost innovation while not neglecting the importance of trusting relationships in which fostering technological transfers appears as more beneficial than keeping innovation limited to one company only (the strength of the network) [INS 01].

As such, we see that the concept of network enterprise exists at various levels. The mere fact of signing an agreement does not suffice for the relationships between two companies to be qualified as a partnership. The "will" to unite, at least in the medium term, into a specialty subcontracting relationship is a necessary marker of the initiation of a partnership. This is in that perspective that we speak of "network enterprise". As for "collaborative enterprise", it is the ultimate stage of the network enterprise. This is where notions of intelligence subcontracting, of formalized agreements etc., will come into play. This involves detecting elements that hint at a negotiated long-term agreement in which a certain degree of maturity and stability emerges from an evolved partnership relationship.

2.2.3.1. *Virtual enterprise*

According to [VER 96], a virtual enterprise is a relatively temporary network of formally independent companies or individuals uniting their means, skills or resources in order to achieve together a project that could have exceeded the capacities of each involved entity if considered alone. A virtual enterprise seeks to exploit volatile opportunities, to gain access to new markets, and to share costs and risks without adding an overly weighty organizational superstructure. To do so, it uses the new possibilities opened by novel information and communication technologies. [PER 98] adds that the virtual enterprise features teams that not only are multi-site and disseminated all over the globe, but are also multilingual, multicultural, multifunction, multitasking and often even multi-shareholders in the case of joint ventures.

According to [LUI 96], a virtual enterprise has four main features:

1) Virtual enterprises are dynamic and they can quickly react to market developments.

2) They tend to be short-lived: although there are examples of virtual enterprises lasting up to eight or nine years; however, most of them only exist for a few months.

3) The hierarchy of virtual enterprises is loosely defined and not a subject of great concern. This is partly due to the fact that hierarchies are not necessary in an ephemeral organization. In a virtual enterprise, the nature of relations rather than official hierarchical ties determines the success of the company. In this type of company, relationships are based on trust.

4) The fundamental skills of a virtual enterprise are spread across the whole group or consortium of groups.

[LUI 96] has listed some reasons to explain why some organizations choose this path:

– *Customers*. Customers are now central to the activity of most companies. Their demands are extremely fickle and volatile, which favor an underlying feature of the virtual company: quick reactivity.

– *Distribution of costs and risks*. Setting up a virtual enterprise enables the sharing of both costs and risks among several partners, some of which are more mature and therefore have a stronger ability to bear the risks involved.

– *Fundamental skills*. Fundamental skills are somewhat paradoxical. On the one hand, companies of all sizes are advised to reduce the number of fundamental skills that define them; on the other hand, companies realize that in order to meet their customers' needs, they may have to acquire novel fundamental skills. The virtual enterprise seems like an appropriate means to dissolve this paradox. It enables a consortium of companies, each of which has quite distinct fundamental skills, to unite and thus offer a full range of fundamental skills.

– *Intellectual capital*. By sharing human resources and therefore their knowledge, expertise and particular transferable skills, companies increase their capital.

– *Control of change, uncertainty and complexity*. This does not really mean that such companies are able to find exhaustive and simple solutions to these very complex issues. It instead means that the people working in these companies are directed to concentrate on the challenges raised by change, uncertainty and unpredictability.

In the automotive sector, the development of the Ford Contour, which Ford wanted to be a universal car (*World Car*), was largely inspired by the virtual enterprise principle. When Ford developed this car, the manufacturer transferred to the market space a key element of its value chain: product development. The design team built and tested prototypes in a computer-simulated environment and shared its plans and data with the partners who were connected to the information network from across the whole world, all the time. In the virtual world of information, the design team established the global manufacturing characteristics and integrated the components and subsystems in a centralized manner while encouraging the providers to partake in the design process. Ford therefore did not unfold the value creation steps on a physical value chain but on a virtual value chain.

We must underline the fact that working within virtual teams requires team members to be especially flexible and able to respect the (cultural, linguistic, behavioral, etc.) differences between members. Team members not only work with four or five teams each day, but they also see their role change according to the various projects they take part in. They may, for example, lead a team in one project and take part as mere support members in another. According to [VER 96], such an organization implies relationships based on trust, a mutual understanding of the manner in which collaborations should be managed, unrestricted sharing of confidential information and, above all, it requires that each member of the team be motivated. Traditional management methodologies also obviously lose all efficiency in this specific type of organization; the new methodologies of "remote management" or "virtual management" require managers to radically change their working methods and learn a flawless mastery of new skills.

Within network enterprises, we can discern two distinct types: virtual enterprises and logistic (or supply) chains. The virtual enterprise, with its relatively temporary nature, induces only superficial changes within each company of the network. The supply chain, on the other hand, implies that each partner truly adopts a new collaborative philosophy, and requires the setting up of a customer-oriented organization with deep changes which are aimed to ensure the long lasting of the collaboration.

2.2.3.2. *Supply chains*

Constant innovation forces companies to focus on immaterial skills, marketing, as well as research and development. The flawless quality that companies aim for requires first to entrust specialists with the production, sales or services, and then to set up robust collaboration processes. This is where information technology tools come into play: they make the various relations between partners of the network concrete and, thanks to eased information exchanges, improve the quality and speed of execution within the organization, thus partaking of global efficiency [DUP 01]. Relieved, the company then shows an increased ability to react to competition and quickly adjust its course. Managing its brand policy also becomes easier. However, it raises issues related to the inertia of the squadron of suppliers, subcontractors, service providers and distributors the new organization involves. This falls under the general issues raised by Sajiv Sidhu: how to detect an event and analyze it in order to make a decision and be able to execute it [BOU 03].

2.2.3.2.1. From the supply chain to its management

The concept of the supply chain was born from the need to set up a form of management geared towards customer service. Increasing both customer service and profitability is therefore not contradictory anymore. Logistics has gone from being a mere cost center to becoming central to a revolution in perspectives. This evolution is reflected by the term "logistics" being replaced by the term "supply chain". More generally, the supply chain can be defined as the set of all activities related to the transformation and circulation of goods and services, from raw material extraction to the end customer [LOH 04]. This definition directly implies the integration of all these activities, be they internal or external to the company. This integration, based on the coordination of activities and processes within the procedure, is usually called "supply chain management". This term encompasses two dimensions: one is a strategic dimension and the other is an operational one. The first dimension focuses on the optimality, for a company, of the chain composition in terms of number of agents and of distributions of roles in the performance of activities. The second dimension focuses on the efficiency of the implementation of the necessary means to coordinate the various functions and activities of the company.

To describe supply chain management from a global perspective is, however, insufficient, as we need to also detail the implementation level. Indeed, according to [BOU 03], five different levels can be identified:

1) To apply fundamental guidelines with the aim of remaining watchful over general quality.

2) To organize in multi-function teams with the aim of offering the best customer service possible.

3) To set up an integrated organization with the aim of managing the business in a more efficient manner, for example by setting up an ERP software.

4) To organize an "extended supply chain" with the aim of creating added value in the company's activity.

5) To create "supply chain communities". Concretely, the company seeks to anticipate the changes of its market by setting up a dedicated organization within which agents are devoted to specific projects. This organization involves permanently or intermittently associating several agents of the company from the levels of design, production, marketing and suppliers.

Throughout these aims, companies can advance from a level where they endure the competition to a level where collaboration enables them to create competitive edges.

2.2.3.2.2. The collaborative mode

Independently from collaborative tools, endeavoring into an approach that involves various professions and skills requires satisfying some conditions of success. For example, the optimization of the system must be performed from a global perspective rather than from the local perspectives of the various independent links of the chain. This global perspective should enable the detection of gaps and weaknesses and the identification of functions where collaboration is relevant. Such collaborations can be effected between services of a single company (intra-company collaboration) or across various partner companies. This mode of functioning also requires some amount of upstream communication in order to establish goals that should be common instead of antagonistic. Finally, working in a trusting and respectful environment is a compulsory requirement.

Most collaborative processes can come into play in the company at the levels of product design, sales prediction, production planning, purchase/procurements, etc. Such collaborative processes are, for example, used during unavoidable operations such as collaborative predicting and planning as well as during the collaboration with suppliers [BOU 03]. The three types of collaboration are detailed below:

– *collaborative predicting*: involves, for example, the setting up of a method of supervision and management of sales prediction which is shared by the various agents within the framework of a common platform. The aim is to set up a central tool to process the information that flows up from field reports;

– *collaborative planning*: collaborative planning is based on tools which allow a good overview of the production capacities. Once the various constraints have been taken into account, collaborative planning enables the design of plans and schedules with a detailed view on the various agents to involve and on the timing of their intervention;

– *collaboration with suppliers*: companies also have a significant need to list "good" suppliers, and an operational common utility to work with such suppliers does enable the massive reduction of purchasing costs.

2.2.3.2.3. Benefits of supply chains

According to [LOH 04], embracing the concepts of supply chain results in the following three types of practical improvements:

– cooperation at the functional level along the whole supply chain allows for shortened delivery times, increased flexibility and improvement in the rapidity with which new products are introduced. Many companies are indeed functionally organized, which means that their organization is centered on production process activities, which in turn makes it difficult to control the supply chain as a whole. In order to provide a better service to customers, companies must, however, synchronize functions such as marketing, sales, distribution, manufacturing and procurement [AND 97, BUZ 95, LAM 96];

– a better synchronization enables a company to overcome functional and national barriers. National boundaries crossing has especially occurred in Europe where many companies moved from one country to another and thus shifted from a strongly national organization, with local production, products and customers to an organization with a more specialized production in which one particular plant produces a specific part of the range of products for the whole of Europe. Sales and marketing have become partly centralized. The supply chain thus needs to be monitored at a continental scale [ABR 97];

– improving operations along the supply chain enables to provide a better service to the end customer, with reduced costs in the supply chain as a whole [JOH 87].

In this context, the phrase "collaborative enterprise" is often used to identify companies working jointly to determine a shared project.

2.2.4. *Virtual organizations and clusters*

Nowadays, the concept of the virtual organization gradually replaces that of the supply chain. Virtual organization is indeed an extension of the supply chain concept, because it includes other types of members, members who need not be companies [BOU 03]. A virtual organization is a set of independent organizations which collaborate with each other by means of specific information networks in order to provide a set of products and services of functionalities as if they were a single organization [LOH 04]. This definition implies that a virtual organization behaves and can be managed as if it was a single company. Virtual organizations, however, have aspects that generate new types of management challenges [GOR 04].

The first attempts of collaboration in the form of network enterprise were too focused on the technological means and underestimated the challenges raised by the virtual organization of the network. The agility and dynamism which the virtual organization requires may be curbed by the need to establish a common operational basis and to build trust between the various partners. Even if flexible enough, logistic infrastructures do exist: building a truly trusting atmosphere and achieving the necessary re-organization of each company raise challenges which sometimes strongly limit the business collaboration process. Establishing trusting relationships between companies is a long and gradual process. Setting up rules of governance or, in a simpler approach, developing common practices can prove extremely difficult, especially when the involved companies have very different cultures. This explains why organizations which form collaborative networks with companies without previous collaboration experience may restrict themselves to simple business collaborations [CAM 08].

Creating clusters of industrial or service companies is an approach that can help overcome such hurdles. Clusters can also ease the quick forming of virtual organizations inspired by temporary common projects. The concept of enterprise cluster, not to be mixed up with that of the virtual organization, represents a business catchment of companies associated with supporting institutions such as state-run public organizations or research institutes. These organizations have the potential and will to cooperate through the setting up of long-term cooperation agreements which define functional principles and common infrastructures. Clusters are usually unified by relationships such as the customer–supplier relation type, by a grouping around a common technology, by common distribution channels or even by a shared theme [CAM 08, FLO 00].

The most frequent case is that of a cluster made up of organizations located in a well-defined geographic area. Geographic proximity indeed undeniably benefits the efficiency of the collaboration because it can, in addition, ease the adaptation to local needs and create a “community” atmosphere. With the advent of more efficient information and communication infrastructures, such long-term associations are not necessarily motivated by geographical proximity anymore. Cultural relations and human relationships also determine the choices involved in the creation of these associations. They constitute the environment in which virtual organizations can be bred. Each time a member of the cluster launches a new project, some companies of the cluster may be chosen to form a virtual organization around that specific project [AND 97].

Another interesting aspect of geography-based clusters consists of the possibility they offer to combine the high standards of various partners who can be organized around various economic models, ranging from digital economy to traditional industries and green economy. The whole becomes a sustainable environment, an ecosystem of local companies which breeds and benefits from the regional assets and cultural features. The cluster can be in charge of the exploitation of local resources and skills through a dynamic and flexible organization of all the partners around each common project. Clusters furthermore gather unique sets of local skills which are perfectly suited to the local culture and customers, which enables efficient cooperation with globalized companies in the offering of products and services which are adjusted to local expectations. Thus, for example, members of an industrial pole can play an important part in the customization and final assembly of products aimed at the local market, even if the basic components may be produced elsewhere. In times of exacerbated competition and of market turbulences, efficient organization and management of local industry or service companies is a promising approach to sustainable regional development. In addition to the aforementioned benefits of cooperating within dynamic virtual organizations, there is the possibility of sharing experience and costs of introducing, for example, new information systems across an industrial cluster [CAM 08].

Some researchers adopt an approach that is more theoretical than practical. They often focus on very precise scenarios which lead them to advocate considering virtual organizations in a totally “open” context and therefore considering clusters as a seemingly restricting approach. The reality of actual companies, however, proves the soundness of the cluster approach we just presented. Furthermore, a great number of real, life-scale experiments and local initiatives have appeared during the last decade across the world, in Europe [BUZ 95, LAM 96], Japan [ABR 97], Brazil [JOH 87, VAR 06], Mexico [CAM 05, FLO 00] and in the USA [CAM 08, GOR 04].

A concrete cluster example consists of the organization of more than 70 active members within the "Virtuelle Fabrik" cluster [AFS 05, PLU 05] in Switzerland and Southern Germany. Advances in information and communication technology now open new avenues to benefit from the potential of this concept, as they provide adequate environments for the rapid breeding of dynamic virtual organizations.

Current trends of mass customization have furthermore highlighted the necessity to take into account the regional preferences, specificities and constraints of the market. The current challenge consists of enabling collaborative innovation which would involve, on the one hand, a network of SMEs (manufacturers, designer, etc.) and on the other hand, customers and an interface between the two. Clusters are thus rapidly evolving to tackle the much more complex issue of tight collaboration with the customer. The aim is to involve the customer in the network and to develop co-innovation with him, as illustrated in Figure 2.4 [AFS 04].

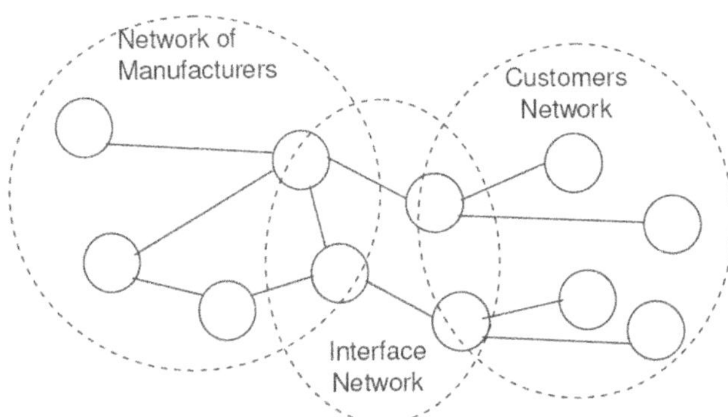

Figure 2.4. *Customer involvement in the network [AFS 04, CAM 07]*

It has been sometimes assumed that partners involved in the creation of a new virtual organization may be easily identified, simply by selecting them from the set of people available within interested companies and/or organizations of the merged collaboration network. In fact, in order for the quick construction of collaboration networks to be successful, the basic unavoidable requirement is that potential partners be ready and willing to take part in such a collaboration. To meet this requirement, we need to prepare, among other things, a common interoperable structure and a shared cooperation agreement. Any collaboration also requires that organizations establish a minimal level of trust. The concept of a breeding environment therefore emerged as a prerequisite to the effective creation of dynamic virtual organizations. Figure 2.5 shows a vision of next-generation clusters and details how the creation of dynamic virtual organizations can be improved with preexisting clusters.

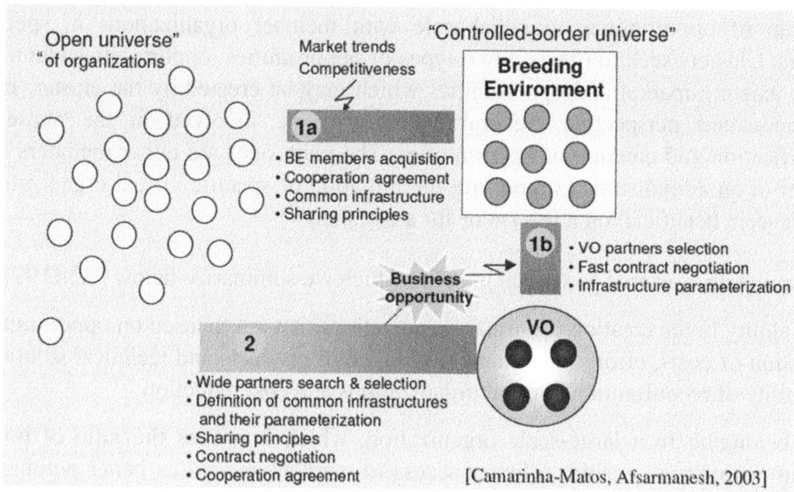

Figure 2.5. *Two approaches to forming virtual organizations [CAM 03]*

The concept of a breeding environment (usually attached to a business sector) emerged as the necessary context of effective creation of dynamic virtual organizations.

Cultural relations and especially human relationships are also important motivational factors which influence the initiation and sustainability of such associations, which are represented by clusters.

Clusters fall into two main categories: regional clusters and global clusters. A regional cluster mainly involves organizations (of various sizes) of a geographical area, whereas a global cluster involves organizations which are spread around the world. In addition, regional and global clusters can either be focused on a specific business sector or be multi-centered [CAM 08].

Clusters usually encourage direct contact between organizations in order to enhance the involvement of each member and to create more collaboration opportunities. Classically, virtual organization breeding environments are set up in a given geographical area and in a given industrial sector. Such focusing enables the emergence of a common enterprise culture, of a feeling of belonging to a community and of a focus on the local domain of specialty. The very structure of clusters, however, nowadays enables the overcoming of regional barriers and the preparation of virtual organization breeding environments of planetary scale.

The main goal of clusters is the efficient creation of virtual organizations. As such, the motivation for creating clusters is mainly based on the identification and creation of opportunities to collaborate with member organizations in specific sectors. Clusters seek to develop two types of opportunities: opportunities identified on an existing market and opportunities which may be created by the cluster, from an innovation perspective. Several agents may be involved in the phase of identification and creation of opportunities. The main ones are either members of a cluster or an administrator supporting the initiation of specific virtual organizations which seem beneficial on a market or for a company.

Setting up clusters has several benefits which we summarize below [AND 97]:

– agility in the creation of virtual organizations that are focused on opportunities: reduction of costs, effort and complexity linked to products and technical solutions, flexibility of reconfigurable virtual organizations and cost reduction;

– belonging to a large-scale organization, which improves the ratio of power during negotiation, enables a better access to markets as well as better purchasing conditions (sharing forces);

– availability of efficient information infrastructures for cluster members, which improves the interoperability and collaboration between members;

– availability of shared information and general knowledge for members (for example, the definition of standard products and procedures), of software, etc.;

– availability of services and information tools that both motivate and ease the configuration and elaboration of virtual organizations: creation of systems to encourage initiatives, reputation services, partner research services, contract negotiation support, etc.;

– proactive management of available skills and resources in the cluster;

– availability of consulting services for cluster members through support institutions: brand management, continued education and training, etc.;

– introduction of methods and procedures to enhance trust between cluster members: for example, records of performance history, definition of trust and solvency criteria;

– elaboration of general guidelines of collaboration: constitution of rules of conduct, work and sharing, of principles and value systems, of ethics and collaborative culture, of intellectual property rights protection methods, etc.;

– improved chance to partake in a virtual organization for each member of the cluster, thanks to the listing of member profiles in a directory with their skills, resources products, services, etc.;

– increased potential and risk-taking capacity for virtual organization planners, thanks to the reduction of configuration efforts, greater variety of available skills and resources, solvency and trust indicators as well as past profitability of the cluster members.

In order for a cluster to appropriately function and breed all these benefits, several requirements must nevertheless be met [AND 97]:

– setting up of a cluster, which must be associated with the setting up of a specific information system providing a package of tools to ease both administration as well as configuration tasks and the creation of new virtual organizations;

– active involvement of member organizations, including disclosure of updated information about their capabilities, resources, costs and planning;

– setting up of a global model for the establishment of clusters, not forgetting issues of financing and market exposure;

– setting up of a management strategy, of governance rules and internal regulations with possible sanctions, about the sharing of both contracts and actual work.

2.2.5. *Virtual communities*

Virtual communities gather people with similar interests in order to communicate, share and exchange information or merely satisfy our need for social belonging. Typical examples include communities focusing on specific sport practices, scientific fields, trades, etc. Virtual communities are created and grow on the Internet. They can, for example, be newsgroups, chats, collaborative work spaces, free encyclopedias, etc. Such communities invent novel types of social interaction, which in turn bring about new behaviors and ways of sharing and creating knowledge [CAM 08, WEN 00].

There also are practical communities; they have been in existence for many years [LEA 00]. They are “informal groups of people sharing some expertise or passion for a common project and who freely share their experience and knowledge. This constitutes new perspectives on knowledge” [FLO 07]. [KAI 04] stresses the fact that practical communities have been growing in numbers and formalization over the past few years, because they develop critical organizational knowledge. Most communities are “entities which are transversal to organizations and have people in charge of finding and sharing the best practices in order to help community members work more efficiently”.

When a practical community uses information networks as well as most of the practices and tools of virtual communities, it becomes a Professional Virtual Community (PVC) [CAM 08].

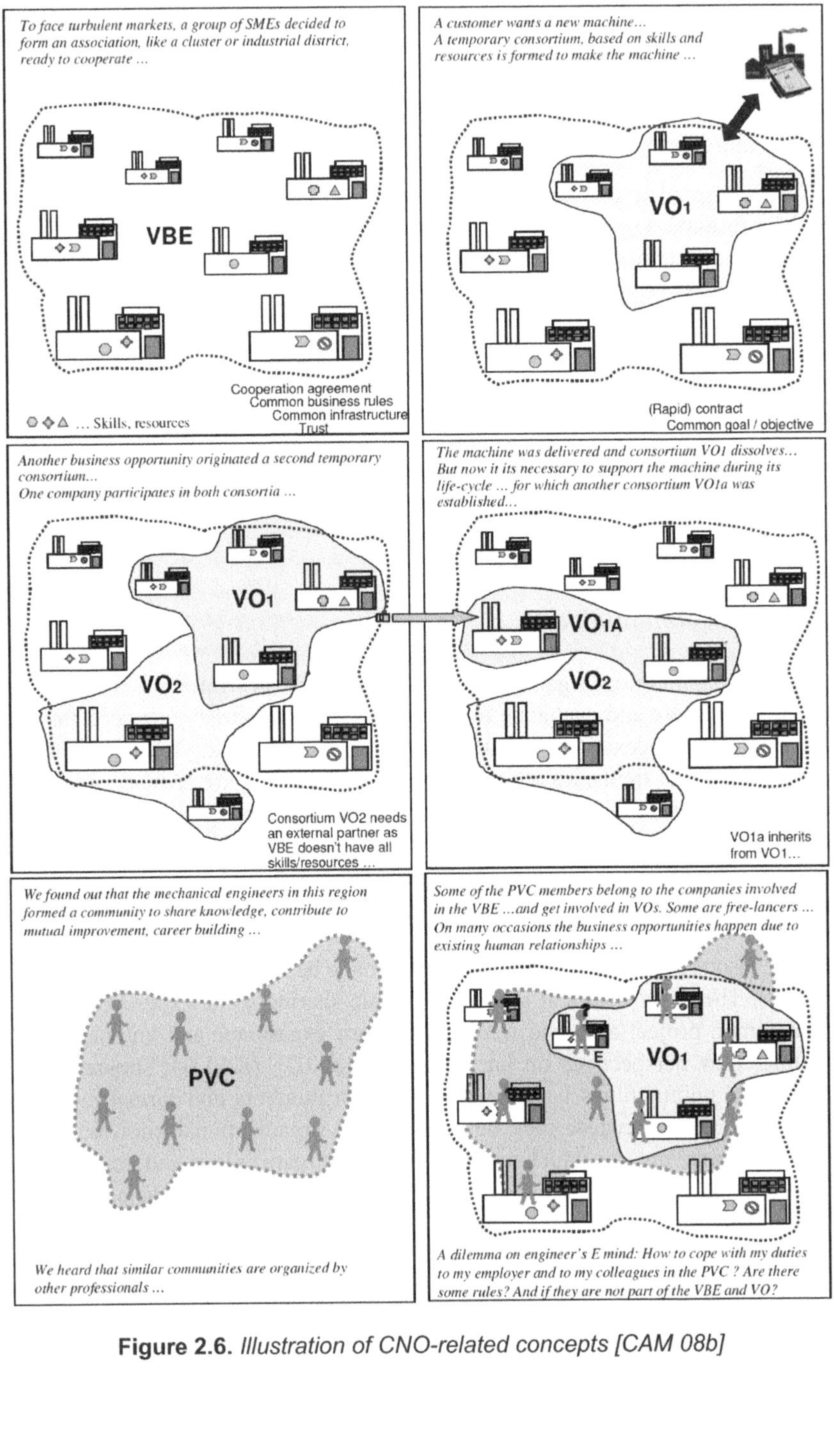

Figure 2.6. *Illustration of CNO-related concepts [CAM 08b]*

A professional virtual community is made up of individuals of similar trades coming together. PVCs ease collaboration between their members and provide environments that breed business opportunities in the same way as clusters favor the creation of virtual organizations.

Virtual communities and practical communities are not new concepts, but they show specific aspects and an increased significance when they are considered from the perspective of collaborative organization networks. The communities, whether they spontaneously emerge, are promoted by companies or entailed by work relations, are attached to specific social rules which can be deducted from the involvement of their members in underlying organizations. Figure 2.6 presents a scenario to illustrate the basic afore mentioned concepts [CAM 08].

They can, for example, occur in the field of collaborative engineering where teams of engineers may be located in various companies but nevertheless collaborate on a common project such as the co-design of a new product or the completion of a consulting mission. This trend is also significant among other professional communities (consultancy communities, for example) who share the whole of the knowledge of their trade, including the cultures attached to specific specialty domains, issue perception, problem resolution techniques, behavioral models, etc.

2.3. Classification of the various types of collaboration and interoperability

Detailing the various types of networked companies enabled us to highlight points for which interoperability issues emerge. However, in order to clarify our discourse, we propose a classification of the various types of collaboration we presented in section 2.2.

This classification is organized along two axes: long-term strategic collaborations and project-oriented collaborations (Figure 2.7).

2.3.1. *Long-term strategic collaboration*

Two different understandings of a long-term strategic collaboration can be identified: virtual organization breeding environments and professional virtual communities. The first type enables the development of an environment favorable to the elaboration of common projects (product creation, services, etc.) and the second one enables exchanges among specialists of a specific domain. The first understanding thus focuses on the ends of the collaboration, which become achievable goals. Achieving these goals requires intense exchanges, encompassing choices of tools and methods. The second understanding solely focuses on exchanging about practices (tools, methods, etc.) without any aim involving common projects.

2.3.1.1. *Project-oriented collaboration*

In this type of collaboration, companies or organizations cooperate around a single project. This means that once the single project is completed, the collaboration is meant to be questioned by the involved parties. They can then choose to pursue it with the same partners on a new project, to further it with a different set of partners or to terminate it.

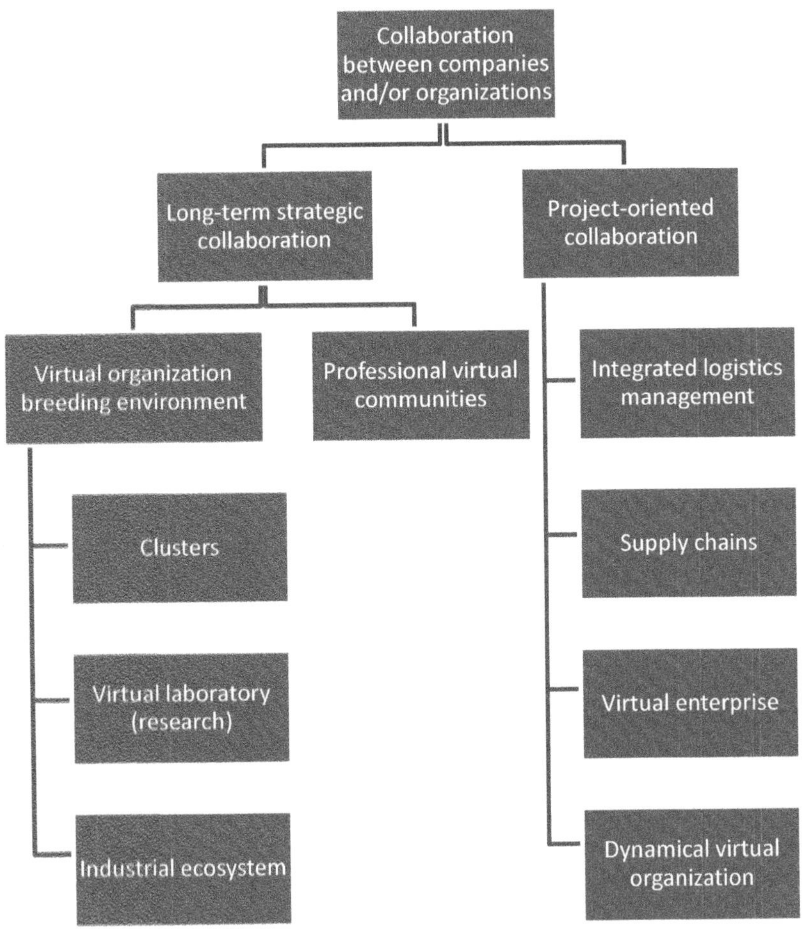

Figure 2.7. *Taxonomy of the various types of collaboration*

In order to define interoperability relative to the types of collaboration between companies or individuals, we used the interoperability framework detailed in the previous chapter. This representation indeed enables the precise identification of the level of interoperability required by each type of collaboration (detailed above, Table 2.1). In the "concern" column, the various categories are ordered from the lowest level of interoperability (data) to the highest (business). Each level of interoperability is encompassed at the immediately superior level. This is why each type of collaboration only appears on one line. We can note that all types of network enterprise require an interoperability whose level is equal or greater than the "service" level. We have chosen not to mention the traditional relationships between companies which, for example, establish customer–supplier type relationships. Since the "data" level of interoperability corresponds to this type of relationship, the corresponding line is empty.

Concern \ Barrier	CONCEPTUAL	TECHNOLOGICAL	ORGANIZATIONAL
BUSINESS	Dynamic virtual organization clusters	Virtual enterprise Dynamic virtual organization clusters	Virtual enterprise Dynamic virtual organization clusters
PROCESS	Supply chain	Professional virtual communities Supply chain	Supply chain
SERVICE			Integrated logistics management
DATA			

Table 2.1. *Framework for the interoperability of companies, applied to the various types of network enterprises*

The framework of interoperability offers a clear and concise perspective on the concerns and barriers to interoperability in relation to the type of collaboration established by the partners. This perspective enables the detection of interoperability issues before even the initiation of the collaboration. It thus enables the elaboration of actions to realize in order for the collaboration to develop in the best possible conditions.

2.4. Conclusion

There are numerous types of collaboration; each of them is suited to specific, distinct needs and situations. This chapter enabled us to classify and to compare them as well as to present their main features. The interoperability needs that they generate are different from each other. Although it would be impossible to detail fine-grained specifications (which should be adapted to specific features of each actually involved company), this chapter classified the needs with the framework detailed in Chapter 1. It thus contributes to providing lines of thought for each specific case of collaboration.

2.5. Bibliography

[ABR 97] ABRAHAMSSON M., BREGE S., "Structural changes in the supply chain", *International Journal of Logistic Management*, vol. 8, no. 1, pp. 35–45, 1997.

[AFS 04] AFSARMANESH H., MARIK V., CAMARINHA-MATOS L.M., "Challenges of collaborative networks in Europe", in *Collaborative Networked Organizations – A Research Agenda for Emerging Business Models*, Springer, Berlin, 2004.

[AFS 05] AFSARMANESH H., CAMARINHA-MATOS L.M., "A framework for management of virtual", in *Collaborative Networks and Their Breeding Environments*, Springer, Berlin, 2005.

[AND 97] ANDERSON E., DAY G.S., RANGAN V.K., "Strategic channel design", *Sloan Management Review*, vol. 38, no. 3, pp. 59–69, 1997.

[BOU 03] BOUCHENAKI N., "Le travail collaboratif à la une des préoccupations", *Pôle Productique Rhône-Alpes*, available at: http://www.productique.org/web/web3.nsf/0/254980E72E861ABFC1256D05002D772C?opendocument, 2003.

[BUZ 95] BUZZEL R.D., ORTMEYER G., "Channel partnerships streamline distribution", *Sloan Management Review*, vol. 36, no. 2, pp. 85–96, 1995.

[CAM 03] CAMARINHA-MATOS L.M., AFSARMANESH H., "Element of a base VE infrastructure", *Computer in Industry*, vol. 51, no. 2, pp. 139–163, June 2003.

[CAM 05] CAMARINHA-MATOS L.M., AFSARMANESH H., OLLUS M., *Virtual Organizations: Systems and Practices*, Springer, Boston, 2005.

[CAM 07] CAMARINHA-MATOS L.M., "Collaborative networked organizations in manufacturing", *Proceedings of IFAC Conference on Cost Effective Automation in Networked Product Development and Manufacturing*, Monterrey, Mexico, vol. 40, no. 19, pp. 187–198, 2–5 October 2007.

[CAM 08a] CAMARINHA-MATOS L.M., AFSARMANESH H., "Classes of collaborative networks", in PUTNIK G.D., CUNHA M.M. (eds), *Encyclopedia of Networked and Virtual Organizations*, vol. 1, Information Science Reference, 2008.

[CAM 08b] CAMARINHA-MATOS L.M., AFSARMANESH H., OLLUS M., *Methods and Tools for Collaborative Networked Organisations*, Springer, Berlin, 2008.

[CHE 93] CHEVALIER J., PONS P., *La logistique intégrée*, Hermès, Paris, 1993.

[COL 96] COLIN J., "La logistique: histoire et perspectives", *Actes du colloque ARFILOG*, 25–26 March 1996.

[DUP 01] DUPIN J.-B., "Organisation: l'entreprise à l'heure du keiretsu japonnais. La gestion de la chaîne logistique: réguler l'écosystème de l'entreprise, 01 informatique", available at: http://www.01net.com/actualites/la-gestion-de-la-chaine-logistique-reguler-lecosysteme-de-lentreprise-167171.html, October 12, 2001.

[ESN 96] ESNAULT L., L'entreprise-Reseau: Une nouvelle frontière pour les managers?, Groupe ESC Lyon, 1996.

[FLO 00] FLORES M., MOLINA A., "Virtual industry clusters: foundation to create virtual enterprises", in AFSARMANESH H., CAMARINHA-MATOS L.M., ERBE H. (eds), *Advanced in Networked Enterprises – Virtual Organizations, Balanced Automation and Systems Integration*, Kluwer Academic Publishers, Berlin, 2000.

[FLO 07] FLORES M., BOER C., HUBER C. *et al.*, "The role of universities developing new collaborative environments; Analysing the Virtuelle Fabrik, Swiss Microtech and the TENET group", in *Establishing the Foundations of Collaborative Networks*, Springer, Berlin, 2007.

[GOR 04] GORANSON T., "Some American research concerns on VO", in *Collaborative Networked Organizations – A Research Agenda for Emerging Business Models*, Springer, Berlin, 2004.

[INS 01] INSTITUT DE RECHERCHE SUR LES PME, La PME transformée, University of Quebec, May 2001.

[JAC 94] JACOB R., RHEAULT M., JULIEN P.A. *et al.*, "L'entreprise partagée et l'approche juste-à-temps", *les cahiers du GREPME*, pp. 94–99, 1994.

[JOH 87] JOHNSON H.T., KAPLAN R.S., *Relavance Lost: The Rise and Fall of Management Accounting*, Harvard Business School Press, 1987.

[JUL 94] JULIEN P., "L'entreprise partagée: contraintes et avantages", *Gestion*, vol. 19, no. 4, pp. 48–58, 1994.

[KAI 04] KAIHARA T., "A challenge towards the Japanese industry: industrial cluster", in *Collaborative Networked Organizations – A Research Agenda for New Business Models*, Springer, Berlin, 2004.

[LAM 96] LAMPEL J., MINTZBERG H., "Customizing customization", *Sloan Management Review*, vol. 37, no. 4, pp. 21–29, 1996.

[LEA 01] LEAVITT P., HASANALI F., LEMONS D., *Building and Sustaining Communities of Practice*, American Productivity & Quality Center (APQC), Houston, 2001.

[LOH 04] LOHMAN C., FORTIN L., WOUTERS M., "Designing a performance measurement system: a case study", *European Journal of Operational Research*, vol. 156, no. 2, pp. 267–286, 16 July 2004.

[LUI 96] LUIK J., "L'entreprise virtuelle", Forum hosted by the Ontario Ministry of Economic Development, Trade and Tourism, 1996.

[PER 98] PERLO A., HILLS C., "Réunir et souder une équipe virtuelle", *L'Expansion Management Review*, pp. 114–119, March 1998.

[PLÜ 05] PLÜSS A., HUBER C., "VirutelleFabrik.ch – a source network for VE in mechatronics", in CAMARINHA-MATOS L.M., AFSARMANESH H., OLLUS M. (eds), *Virtual Organizations Systems and* Practices, Springer, Berlin, 2005.

[PRO 96] PROBST A.R., "Vers des systèmes d'information génériques pour les entreprises virtuelles", *2ème colloque international de management des réseaux d'entreprise*, École des HEC, University of Lausanne, 1996.

[VAR 06] VARGAS R., WOLF P., "Virtual collaboration in the Brazilian mould and die making industry", in KAZI A.S., WOLF P. (eds), *Real-Life Knowledge Management: Lessons from the Field*, Knowledge Board, 2006.

[WEN 00] WENGER E.C., SNYDER W.M., *Communities of Practice: The Organizational Frontier*, Harvard Business Review, 2000.

3

Designing Natively Interoperable Complex Systems: An Interface Design Pattern Proposal

Systems Engineering (SE) is a standardized methodological and interdisciplinary approach that has been successfully applied in various industrial fields. This approach serves to support, guide and manage design activities dedicated to complex systems by taking into account stakeholders' advice and expectations throughout the system's life cycle. This chapter focuses on the interoperability feature which has to be seen as a specific requirement in its own right. Indeed, any system needs to interact with other systems, thereby forming a heterogeneous environment. All these systems have not necessarily been designed to function properly and efficiently with one another, whether from a conceptual, technical, behavioral or organizational standpoint. This chapter highlights what seems to be relevant in terms of conceptual definitions and modeling framework whenever a (group) of engineer(s) intends to design what we call here a "natively interoperable system" or at least a system maximizing its abilities to be interoperable. To proceed, a clear definition of the concept of interoperability applied to complex systems is the first prerequisite. The second prerequisite consists of establishing the needs of a design team assigned to design such a system. An interface design pattern model with sufficient generic, formal and pragmatic qualities is proposed.

3.1. Introduction

Systems Engineering (SE) is "*an interdisciplinary approach and means to enable the realization of successful systems* [socio-technical or technical]. *It focuses on defining customer needs and required functionalities early in the development*

Chapter written by Vincent CHAPURLAT and Nicolas DACLIN.

cycle, documenting requirements, and then proceeding with design synthesis and system validation while considering the complete problem" [INC 03, FIO 12, BKC 14, BLA 11]. SE has been successfully applied for a long time in various fields, e.g. aeronautics [RGA 05, NAS 07], transportation [CAL 07], health care [KOP 07], manufacturing or defense. Based on system thinking and system sciences principles [LEM 94, FÉL 07], it promotes at the same time the use of models and modeling languages, e.g. [MOD 10] and [SYS 14a] and introduces the notion of Model-Based Systems Engineering (MBSE) [INC 08] and a processes-oriented approach [EIA 98, ISO 08, IEE 05, BKC 14]. It takes into consideration frameworks, e.g. DoDAF [DOD 07] and is supported by tools, e.g. [COR 14], [INC 14], [MOT 14], [SYS 14b], [SA 14] and [CRA 14], for overseeing and managing the design of a system of interest (SOI), "*the system whose life cycle is under consideration*". Numerous research and development works have focused on interoperability in recent decades, but remain poorly applied in the SE domain. This is, however, an essential feature of any kind of SOI to maximize and maintain its interoperability, i.e. its ability to *interact efficiently with other systems in various situations, throughout its life cycle* or *to exchange and share contents of flows (items carried out by the flows) and work together without unwanted effects*. Indeed, many examples from industry and from the application fields mentioned above have highlighted that problems related to a lack of interoperability lead to delays, failures, dysfunctions or shortcomings all along the system's life cycle; problems that would have been much more manageable if they had been detected earlier in the design stage. More generally, studying a system's interoperability requires increasing its potential to be used in various contexts, more or less unpredictable. In this case, an interoperable system can be considered as independent of such contexts. As a result, improving the interoperability of a given system can be considered as an alternative objective to improve the system's ability to be definitively integrated into a given and completely identified upper-level system.

This chapter aims to provide methodological elements in order to enrich the SE approach. They are dedicated to support in the design of complex systems that may maximize their ability to become interoperable or even to become natively interoperable. The first aim is to establish the intuitive conceptual hypotheses on which these contributions are based and particularly the strong dependency between the notions of interoperability, interaction, effect and interface. The second aim consists of enhancing the traditional requirements' checklists used in the SE field by defining a category of interoperability requirements. Third, it proposes interaction/effect modeling principles and an interface design pattern model for facilitating interfaces design. This will support verification mechanisms, not presented in this chapter and inspired from [CHA 13a] and [CHA 13b], within a coherent and shared collaborative SE workspace.

3.2. Work program: context, problematic, hypothesis and expected contributions

The design of natively interoperable systems remains difficult due to many conceptual, technical and business limitations. First, two systems may have to interact and exchange flows to fulfill their mission. Conversely, two systems can also interact in a undesired but unavoidable manner, for instance in the military field when the behavior of one system implies a modification of the mission devoted to another one. In both cases, these *interactions*, desired or not, may induce *effects* from various types. Subsequently, they require defining and using *interfaces*. Although the design of an interface is a well-known problem in SE, these notions of interaction and effects still remain neglected in the guiding of the design stage. Second, SE standards promote lists of functional and non-functional *requirements* that limit *interoperability* definition despite the fact that technical (associated with a system's functional and behavioral vision), structural and organizational aspects of this system are simultaneously concerned. This is a limitation, particularly, when considering large systems or system of systems for which [FER 06] introduces the notion of design for interoperability. In the same way, the design of an interface requires modeling means that seem too basic and insufficiently formalized to take into account the interoperability problem. It remains difficult to assess the real added value of the interface and to detect and analyze the effects mentioned above that are not necessarily observable without a descriptive model, turning out to be, at times deleterious. As evidence, SE tools emphasize that modeling languages do not integrate these concepts and relationships, despite being necessary to consider the interoperability problem fully. Third, engineers consider that achieving interoperability consists simply of exchanging flows with other systems within all specified situations. This can be done by using one or both of the two approaches, i.e. by adhering to published interface standards or by making use of a "broker" of services able to assume interface role between these systems, possibly, "on the fly". These practices limit again and crucially the analysis of the real expected system interoperability, for instance, in avoiding unexpected or reverse effects that the interface is unable to prevent or, failing this, to protect the system itself. Researchers (e.g. [MAL 10, MAL 11]) have proposed the means for identifying, characterizing and resolving interoperability problems and barriers. For instance, within the corporate field, several interoperability maturity models [CLA 99] and frameworks [NAU 08] serve to evaluate level of interoperability of an enterprise [DAC 08]. We feel that some of these research results may be applied or adapted to the complex socio-technical systems targeted by SE projects. Finally, a SOI can be identified as shown in Figure 3.1:

– The SOI contributes to an identified, more extensive and more complex upper-level system S and will be more or less integrated in S. In this case, interoperability is not really relevant. As an example, a propulsion system is dedicated and designed to be finally integrated into the airplane. SE helps designers to identify what are the expected interactions and requirements and the design of, among others, the interfaces required between this propulsion system and other subsystems composing the airplane: airframe system, flight control system, etc.

– The SOI must work in tandem with other systems that cannot be definitively identified as an operating environment. For instance, the airplane must be able to use infrastructures and services offered by any airport in the world in conformance with several rules and policies. It has also to be able to operate flights in conjunction with other aircrafts and so on. In this case, designers are also supported by the SE principles; classically, they design interfaces required to meet standards that enable the SOI to fulfill its mission, i.e. by providing all possible interactions with other subsystems ahead of time and in whatever situation in a moving environment. In this sense, we move here from design oriented by integration objectives to design oriented by interoperability objectives.

– The SOI is considered as a System of Systems (SoS), for instance, an Air Transport System. It is an arrangement of systems and components that allows for the achievement of a bigger and more interesting function that cannot be achieved by any of the component systems alone [MAI 94, BIL 14, BIL 15]. An SoS results from assembling existing systems and components, sometimes very quickly, for a more or less short duration, even under emergency conditions. However, such systems and components are not necessarily designed for this purpose and then interoperability of each of these ones become crucial.

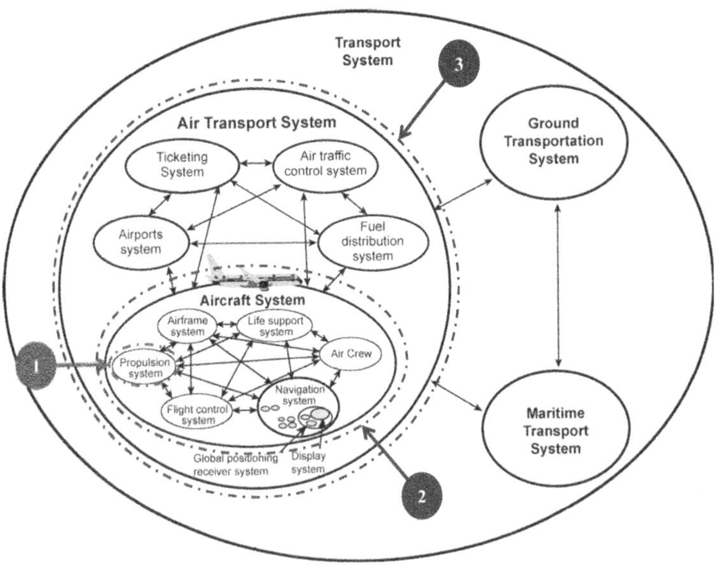

Figure 3.1. *SOI is (1) a system part, (2) a system or (3) a system of systems [BLA 11]*

So we propose, first, a consensual definition of system interoperability and, obviously, of a natively interoperable complex system. Second, after adopting a set of pre-requisites conceptual definitions, we define the notions of interaction, effect, interface and interoperability requirement that enrich current requirement repositories used in industry. The final result consists of proposing an interface model pattern providing concepts and mechanisms that allow engineers to model an interface and then to confirm – by simulation or by checking formally the interface model as proposed in [BER 01] or [MAL 11] by using tools such as proposed in [VVT 11], [DAS 10] and [FOR 11] – whether the considered system can avoid adverse effects due to its interaction with other systems under all specified conditions via its interfaces.

3.3. Concepts

– *System interoperability*: the definition of interoperability depends on the application domain[1]. The goal here is to propose a consensual definition of *system interoperability* for adoption in this chapter and in the SE field. Classically, interoperability involves "*connecting people, data and diverse systems. The term can be defined either technically or comprehensively, in taking into account social, political and organizational factors*". Also "*two or more devices are said interoperable if, under a given set of conditions, the devices are able to successfully establish, sustain and, if necessary, break a link while maintaining a certain level of performance*". In technical systems, interoperability is "*a property of a product or system, whose interfaces are completely understood, to work with other products or systems, present or future, without any restricted access or implementation*". In socio-technical systems, it is defined as "*a property referring to the ability of diverse systems and organizations to work together*", i.e. to interoperate. For instance, enterprise interoperability is defined as "*a cooperative arrangement established between public and/or commercial entities (authorities, parking facility operators, etc.), wherein tags issued by one entity will be accepted at facilities belonging to all*

1 The reader can find various interoperability definitions used in this section in glossaries available on (all last visited and checked 2011-04-12): http://dli.grainger.uiuc.edu/glossary.htm, http://www.eu-share.org/glossary.html, www.csa.com/discoveryguides/scholarship/gloss.php, www.naccho.org/topics/infrastructure/informatics/glossary.cfm, ec.europa.eu/transport/inland/glossary_en.htm, www.nato.int/docu/logi-en/1997/defini.htm, www.cs.cornell.edu/wya/DigLib/MS1999/Glossary.html, www.ibtta.org/Information/content.cfm, dli.grainger.uiuc.edu/glossary.htm, cloud-standards.org/wiki/index.php, en.wikipedia.org/wiki/Interoperability, wordnetweb.princeton.edu/perl/webwn, www.anzlic.org.au/glossary_terms.html

other entities without degradation in service performance". In the same manner, interoperability is considered as "*the ability of a system or product to work with other systems or products without special effort furnished on the part of the customer*". In the military field, NATO also defines interoperability as "*the ability of systems, units or forces to provide services to and accept services from other systems, units or forces and to use these services so exchanged to enable them to operate efficiently together*". Lastly, in transportation systems, interoperability seems to be achieved when "*a transport network [is suitable] for movements without breaking bulk*". Thus, we propose defining system *S* interoperability as:

"*The set of abilities and associated capabilities of a system (namely 'S' from now) that allow S to be and to stay able to exchange and work harmoniously with other systems from its upper level all along its life cycle*:

– *to fulfill a common mission (i.e. the main function for which the overall system is designed), possibly time bounded, while remaining able to perform its own mission and to reach its own objectives through the use of exchanged items with other systems then when S is interacting with these systems whatever may be their nature*;

– *in all specified operational situations (e.g. nominal functioning mode, or functioning modes when facing a risky situation) met throughout its life cycle*;

– *reflecting the stakeholders' requirements under every specified situation.*

This capability indicates and allows assessing before, during or after the interaction – and when placed in its environment – that S does not require or result in major changes to its operations, structure or behavior; consequently, its functional and non-functional requirements (performance, security, safety, ergonomics, human factors, etc.) are not altered. Moreover, this does not induce undue adverse effects (dysfunctions, risks) when S is achieving its mission independently of every other system".

As a result, a *natively interoperable system* can be defined as "*a system designed to maximize its ability to interoperate all along its life-cycle*".

– *Pre-requisites conceptual definitions*: let us briefly recall a set of concepts from the literature on system sciences [LEM 94, JEA 97, DOD 01] and some theoretical foundations of Systems Engineering [BLA 11]. A *processor* aims to transform *items* (of an *information*, *energy* or *material* nature) transported by input *flows*, into new items transported by output flows, under the control of other flows and by using *resources* that support or are involved in processor functioning. For example, a processor can describe a system S, but also a function, an activity or a process when considering the functional *view* of S; moreover, it can also describe a component or an organizational unit involved in S when considering a physical view (organic, organizational). Let us introduce here the so-called "*processor modeling language*" proposed in the SAGACE approach [JEA 97]. It allows to describe a processor in unambiguous terms. Figure 3.2 shows some equivalent notations with modeling languages used in the SE domain such as eFFBD (enhanced Functional Flows Block Diagram) [SEI 09], PBD (Physical Block Diagram), BPMN 2.0 (Business Processes Modeling Notation) and Activity or Internal Block Diagram from SysML [SYS 14a]. So, all of the proposed definitions discussed in the next section can be applied independently of the adopted modeling language. This step offers the designer the freedom of choosing the modeling language for interface design.

The processor behavior described by its *transfer function* details the transformation provided by the processor. This transformation may be described by modeling the modification induced by the processor on one or more *attributes* (formally *Space Shape Time* so-called *SST attributes* in the next) that characterize each item transported by input flows, so as to obtain new or modified items transported by output flows. These attributes concern *space* (e.g. type, definition domain, instantaneous value/default value, etc.), *shape* (e.g. optical, electromagnetic, signal, binary, or linked to the aspect of the pointed out item if it can be considered as dependent from one or several of five senses) and *time* (update frequency, maximum life cycle before updating, etc.) characteristics (Figure 3.3).

The *added value* of the processor *transfer function* can be analyzed by taking into account the variation of the values of the considered items attributes as well as the type of processor (e.g. a function, component, business unit or actor). Analytical methods can be applied to (1) assess processor performance (e.g. in terms of costs, QoS or response times) and (2) check some of the functional and/or non-functional requirements (e.g. by evaluating some of the "*-ilities*"[2]) allocated to the processor.

– *Interaction*: a *relation*, whether planned or desired, or alternatively unwanted or unintentional, between processor P_1 (origin of the relation, hence P_1 plays the *emitter role*) and one or more processors $\{P_2, \ldots, P_n\}$ denoted as $\{P_i\}$ in the next (relation targets, hence they play the *receiver role*). An interaction is defined when:

- there is a flow carrying items of the same type (material, energy or information), or a functional flow of service between P_1 and $\{P_i\}$ (often modeled by an *influence*, *coupling* or *transaction* type of exchange [JEA 97]);

- there is a physical energy field F from thermal, mechanical, pressure, biological or other nature. A list of common fields is given in [MAN 02].

2 "*-ilities*" are defined in [WIK 14] and [WEC 12] as *the developmental, operational and support requirements a program must address (e.g. availability, maintainability, vulnerability, reliability, etc.)*, i.e. non-functional requirements.

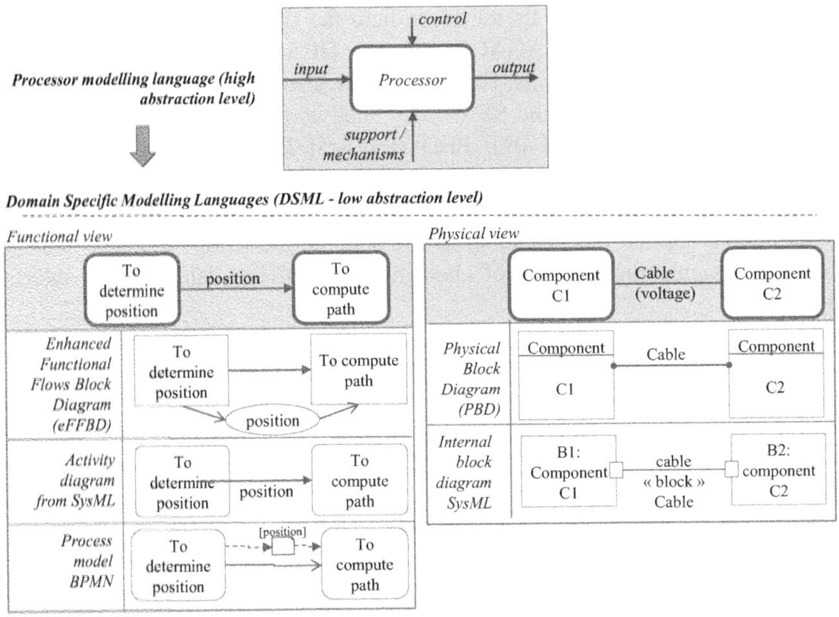

Figure 3.2. *Equivalent notations between some modeling languages used in the SE context*

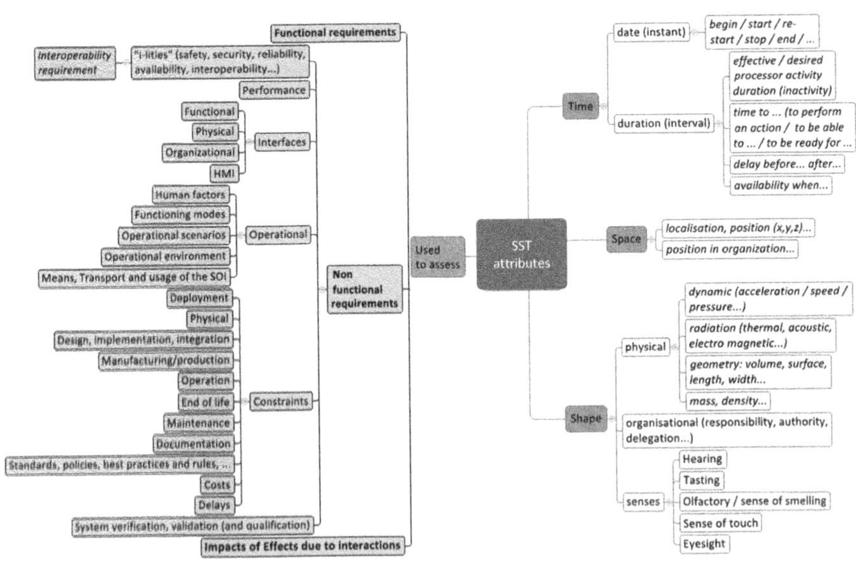

Figure 3.3. *SST attributes categories and examples in use*

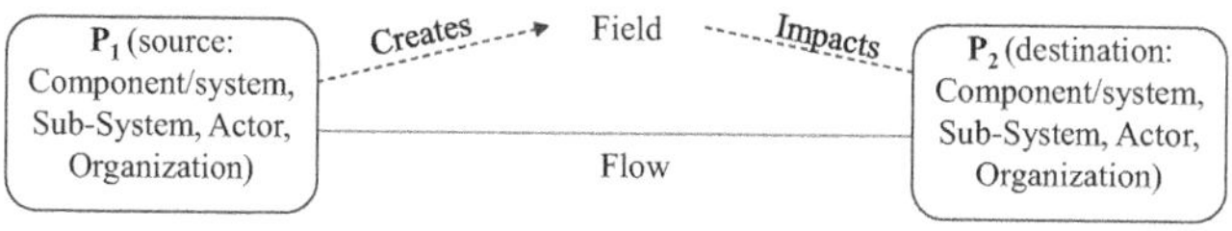

Figure 3.4. *Interaction definition*

– *Effect*: an interaction may induce one or more *effect* defined then as the outcome of the interaction between a processor P_1 and another processor P_i (i = 2, …, n), if the interaction:

- induces one or more P_1 (reverse effect) or P_i (direct effect) TSS attributes variation, *and/or*;

- impacts the satisfaction of one or more requirements (functional and non-functional, i.e. -ilities) of P_1 (reverse effect) or P_i (direct effect), *and/or*;

- impacts functioning, decision-making process, structure or organization of P_1 (reverse effect) or P_i (direct effect).

The effect is characterized in one of various ways, as proposed in [MAN 02]:

– harmful: at least one characteristic of P_i becomes inconsistent with the conditions necessary for its survival. In this case, the pointed out relation causes the emergence of behaviors or physical phenomenon that are often inappropriate, such as resonances, electromagnetic interferences and thermal effects, thereby inducing rather risky situations (accident, incident or malfunction) or damage to operational modes at the source and destination(s). They have to be avoided or simply modified;

– desired but absent: the effect should exist but proves to be impossible. Some non-functional requirements concerning P_1 and P_i have not been verified (performance, safety, security, etc.);

– desired and present: the effect exists and moreover is considered necessary. All requirements concerning P_1 and P_i remain verified;

– appropriate but insufficient or excessive: the effect exists, yet some non-functional requirements concerning P_i are not checked (performance, safety, security, etc.).

The relation is provided by P_1 whether intentionally or not. The effect can be derived from various dimensions, depending on the technical or socio-technical nature of processors P_1 and P_2. According to the case, designers can characterize these effects by using the substance-field model originally proposed in [ALT 12]. The resulting proposed effect model is defined in Figure 3.5.

The database of accessible effects [OXF 12] can help designers first to identify solutions that correspond to problems at the origin of the effect. It allows then to define interface objectives, i.e. to authorize and govern any interaction as defined in the next.

– *Interface*: System Engineering processes shall "[...] *ensure the compatibility, interoperability and integration of all functional and physical interfaces and then ensure that system definition and design reflect the requirements for all system elements: hardware, software, facilities, people and data*" [DOD 01]. As such, an interface is defined in [AFI 06] as "*a boundary across which two independent systems meet and act on or communicate with each other*", i.e. interacts with each other. In detail, an interface is "*the common logical and/or physical border between two or more components (or between the system and its environment), at which the rules of exchange, compatibility, integrity and non-regression are to be respected throughout the system's life cycle*".

Conventional effect model graphical representation	*Effect interpretation*	*Interface objectives*	*Interface mission*
P_i P_j	Harmful	must be detected and avoided	To protect
P_i P_j	Good and desired	must be checked and preserved	To improve exchanging and sharing flows
P_i P_j	Desired but absent	must be created and preserved	
P_i P_j	Excessive	must be reduced and preserved	
P_i P_j	Insufficient	must be improved and preserved	

Figure 3.5. *Effect model principles inducing interface objectives and mission*

From a theoretical point of view, an interface is requested to allow P_1 and each P_i (i = 2, …, n) (Figure 3.5) to:

- *exchange* the requested flow. In due course, it must consider (in the *receiver role*) or make available (in the *emitter role*) the items and then the flows. Among other abilities that depend on its type (functional, logical, physical or organizational processor), this set-up must:

- provide functional skills: emit, receive, transport, adapt (e.g. convert the input format of the exchanged flow), separate, protect, authorize the relation, involve another processor, manage the items (e.g. store, retrieve), etc.,

- respect overall system requirements and, in particular, interoperability requirements as defined below;

- *protect* themselves from, or avoid, the potential effects induced by the relation, or, at least, be able to contain the inappropriate effects within acceptable limits. To achieve this protection, its behavior has to be adapted and/or a set of protection mechanisms or barriers are to be designed to limit risky situations, for instance inspired by resilience engineering principles [HOL 11].

For instance, [US 00] promotes a System Interface Description (SID, Figure 3.6) that highlights various processors nature (node, system, component) having to be interfaced considering conceptual levels of details called here perspectives.

– *Interoperability requirements*: [INC 08] proposes guiding the stakeholder requirement definition [NUS 00] by the use of a conventional checklist in which functional requirements (i.e. "*what must the system do?*") and non-functional requirements ("*what are the* system's *expected characteristics* of performance, '*-ilities*' and constraints *supposed to do*?") are separated. Interoperability might concern functional as well as non-functional aspects of S. However, before assisting designers to design a *natively interoperable system*, it is first necessary to formalize the notion of interoperability requirement.

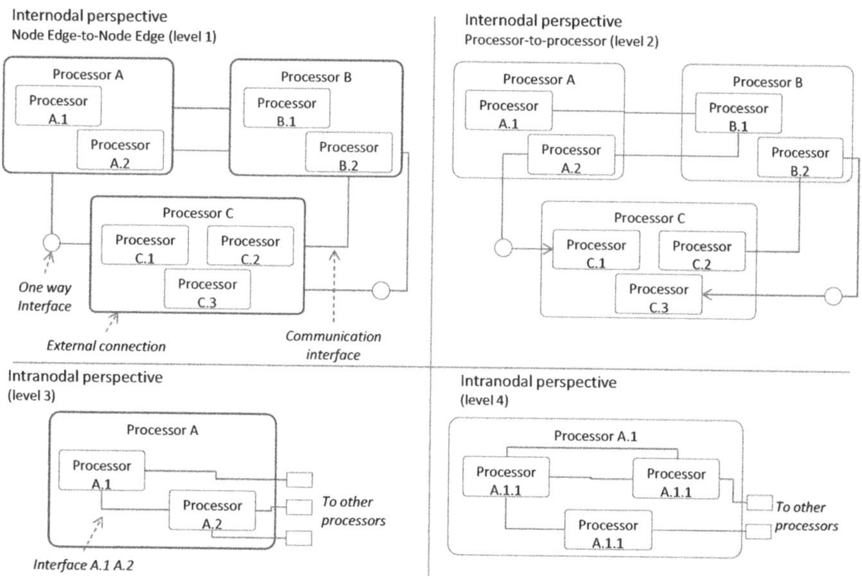

Figure 3.6. *An interface is required for various perspectives and types of processors (inspired from [US 00])*

Inspired by [MAL 10] (applied to collaborative processes seen as being complex socio-technical systems), the interoperability of S is achieved by means of:

– *Compatibility*: S can send and receive flows from other systems in its environment whenever such interactions are needed. This ability tends to be driven by respecting technical standards, communication protocols for technical compatibility or organizational rules and policies for organizational compatibility, described respectively as *technical compatibility requirements*, e.g. required frequency of the exchange and *organizational compatibility requirements*. These must be recognized by all systems having to interact with S.

– *Interoperation*: S operates seamlessly with the other systems in its environment by taking into account flows content being exchanged in order to fulfill its mission; moreover, it has the ability to control, adapt or anticipate problems promptly. S can also influence, and not necessarily intentionally, other systems through both desirable and adverse effects. In this case, the term *interoperation requirements* could be referenced, e.g. lifetime of any item transported before its obsolescence in taking into account states and modes of operation at the origin and relation target (ready, stop, etc.).

– *Autonomy*: S is independent of other system operations and behavior. Autonomy may be decomposed into decisional autonomy (where S assumes its governance and remains capable of deciding actions) and operational autonomy (where S remains capable of preserving its performance in terms of cost, schedule and quality of service). At this point, it becomes necessary to consider *decisional autonomy requirements* and *operational autonomy requirements*.

– *Reversibility*: The relationships between SOI and the other systems are completely reversible, i.e. S can return to an identified configuration or state without causing any problems (dysfunctions, loss of performance, requirement violations, etc.) requiring difficult to manage changes once S no longer needs to exchange with the other systems in its environment. *Relationship reversibility requirement* is the term introduced here.

In conclusion, a system can be then defined as a processor from high abstraction level decomposable in "*a set of elements [of a heterogeneous nature] that accomplish [by interacting with other processors] a defined objective*":

– the processor can be described by building various models using the modeling languages, as stated above. These models allow respectively focusing and representing its requirements, functional and physical architectures, behavioral characteristics (operational scenarios, operating modes) and properties (attributes of time/space/shape, characterizing both static [MIC 08, QAM 12] and dynamic aspects, i.e. temporal or temporized, business rules and axiomatic knowledge as proposed in [CHA 13b] and [CHA 15]);

– the processor evolves in a context called its upper-level system, considered itself as a processor and made of other systems or else components/organizational units or actors;

– the processor must fulfill its own purpose, achieve its own objectives, yet at the same time assumes a role when having to interact (collaborate, exchange, protect, etc.) with other systems in order to achieve the upper-level system objectives.

A causality rule exists whereby: "*A (sub)processor A will be interoperable with (sub)processor B if all elements that compose (from different sources) or refine (from the same source) A and are involved in the relationship with B do not cause interoperability problems, i.e. A and B respect interoperability requirements regarding their own role and objectives within the relation*".

3.4. Interface design pattern model

Following previous propositions and definitions, an interface is then conceptualized as a processor P intending to establish a connection between a processor P_1 (system, component, subsystem, business unit or actor) with its environment composed of a set of processors $\{P_2\}$ in order to (objective 1) transport flows between P_1 and $\{P_2\}$ (or vice versa) and/or (objective 2) protect from, in the sense of avoiding, unwanted effects between P_1 and (at least) one of the $\{P_2\}$ processors resulting from relationship implementation. When considering interface design and following system design principles (e.g. as described for user interface design in [THI 02]), three cases must be raised:

– designing a native interoperable processor P induces the design of each needed or potential interface, by then considering each interface as a sub-processor of P;

– improving the interoperability of an existing processor P can induce global or partial re-engineering of each of its interfaces, considering that each processor P_1 found to play the role of interface can then either replace one of the parts of P or else be added to P;

– improving the interoperability of processor P, by considering P impossible to modify (e.g. P must be definitively integrated into a more complex processor P_1, perhaps assumed to be the upper-level system). It induces the design and addition of new required processor P' playing the role of interface between P and the other processors $\{P_2\}$.

Lastly, designing an interface obviously requires being able to take into account:

– its type: logical or functional at first, it then becomes physical, HMI or organizational;

– the nature and objectives of the interaction and the potential effects it induces on the interfaced processors or on their environment;

– the nature (matter, energy, information) and characteristics (e.g. throughput, processing time, accounting time, space and shape attributes) of exchanged items carried out by the flows;

– interface requirements, including interoperability requirements, features and expected services and characteristics (its own "*-ilities*").

It is also important to determine the expected level of quality of the model, which, in its turn, requires the possibilities and mechanisms of automated verification and, if possible, validation. In accordance with the basic principles of the SE domain, an interface can be viewed as a processor characterized by:

– *Interface purpose*: the interface (objective 1) "*permits ensuring the exchange of flows between two or more systems (components/functions/actors or business units having to be identified), as expected from an efficient (in terms of resources used) and effective (with positive results) procedure*" or (objective 2) "*contributes to improving the protection of a system (to be identified) from an efficient and effective procedure, in taking into account other systems with which the interaction is not mandatory or even inevitable*".

– *Interface mission*: an interface (objective 1): "*ensures the requested interaction*" or (objective 2) "*limits the effects of the interaction between identified processors (systems/components/functions/actors or business units) from an efficient (in terms of resources used) and effective (with positive results) procedure*".

– *Interface objectives*: an interface must respect overall functional and non-functional requirements (including interoperability, but also for instance performance, ergonomics, constraints or verification requirements). They are induced or come from the identified processors being in relation. At least, an interface must improve the identified processors interoperability (i.e. compatibility, interoperation, autonomy and reversibility), reduce effects (having to be detected, identified and then modeled by an effect model) following the interface objectives (as synthetized in Figure 3.5), interchangeable, reliable, affordable, scalable, manageable and interchangeable.

– *Interface typology*: it is proposed to distinguish the functional or logical interface, from physical, human–machine and organizational interfaces as follows:

– *Functional interface*: between functional entities (e.g. functions from functional architecture). The designer creates functional interfaces between functions that model the flows to be exchanged (data, material, energy) in the role of input, output, control (trigger) or resource flows. This notion of function requires determining:

- types of carried out items, contents, origin (external of the SOI or internal) and respective roles in the system;

- whether or not a communication protocol and exchange is required and formalized;

- whether a treatment protocol or flow adjustment is needed taking into account interoperability requirements;

- whether a synchronization protocol or source and target adjustment is necessary.

After allocating functions to the processors, the functional interface evolves into the physical, human–machine and/or organizational interfaces, which are then required between SOI and its context, or else between subsystems and components.

– *Physical interface* [WHE 10]: between the SOI and its context or components or subsystems of an organic architecture [THI 02]. These interfaces are required to:

- enable operating functions on physical flows and hence meeting the functional requirements. For instance, [AFI 06] distinguished five types of physical interfaces:

- spatial: related to physical adjacency for alignment, orientation, serviceability, assembly or weight,

- structural: related to load transfer or content,

- material: related to the transfer of airflow, oil, fuel or water,

- energy: related to the transfer of heat, vibration, electric or noise energy,

- information: related to the transfer of signals or controls;

- respect non-functional requirements (performance, "*-ilities*" such as interoperability when considering non-functional aspect of interoperability, and abilities, e.g. emission, reception or transport of a flow).

– *Human–machine interface*: the activities required for user interface design are already detailed for instance in [GRU 11].

– *Organizational interface*: these interfaces are required between actors and organizational units involved in and required to play roles in the SOI. Exchanges become necessary in conducting sharing, collaboration, communication and cooperation when performing activities to: produce/manufacture, deliver, store, sell, buy, design, manage, control, verify, plan, teach and organize training periods for stakeholders, qualify actors' profiles, decide, etc. These interfaces can be modeled as a collaborative working process model or a virtual organization model for instance.

– *Interface SST attributes*: the goal is here to define what are the requested SST attributes of the requested interface for instance in terms of potential physical elements that can be used to implement the interface (communication components, connections, ports, links, etc.) as shown in Figure 3.3 and such as:

- time, e.g. duration for connection/disconnection, maximum delay before updating value or life cycle duration before obsolescence of the carried out items;

- shape, e.g. dimensions (L*H*D), geometry, weight, radiation from various nature (see the list of possible fields in the effect model);

- space, e.g. position, speed, transfer speed.

– *Interface functioning modes*: as any component, an interface evolves all along its life cycle by passing from a state, i.e. a functioning mode to another one. In each mode, specific functions, SST attributes or requirements have to be highlighted and studied. We propose here to use the classification of operational modes [CHA 13a] as a checklist (Figure 3.7). This allows the designer to select the appropriate modes to be studied and to specify what are the transitions between these modes to allow passing from a mode to the next one. This condition is then computed taking into account external or internal events from the interface, and/or SST attribute variation. Particular scenarios have to be modeled in order to detect and formalize what are the requested functions in each mode.

SoI life cycle Phases			*Generic Operating Modes (concern an instance of the expected SoI)*
System Realization		R	Producing SoI exemplar: integration / verification / transition / validation / qualification
System exploitation	System Deployment	D1	SoI is waiting for deployment on operational site
		D2	SoI is removed from operational site
		D3	SoI functions for tests, maintenance, or training out of any operational site (off-site)
	System Operations	O1	SoI is deployed (Transport / Assembling / Installation / Qualified) on operational site (on-site) and ready for operation
		O2	SoI is prepared to assume its mission in nominal situation and conditions
		O3	SoI functions in normal / nominal situation and conditions
		O4	SoI is prepared to end normally its mission
		O5	SoI functions for tests, maintenance, or training on operational site (on-site)
	System Dysfunctions Management	DM1	SoI stops after a default or dysfunction detection (from internal or external cause)
		DM2	Dysfunction or failure diagnosis (causes / effects / maintenance decision)
		DM3	SoI functions in non-nominal situation (degraded behaviour but fulfils its mission with lower efficiency)
	System Maintenance and Evolution	ME1	Diagnosis and Corrective Maintenance
		ME2	Diagnosis and Preventive Maintenance
		ME3	Diagnosis and Adaptive Maintenance
		ME4	Diagnosis and Maintenance for Evolution (request partial re engineering for new functionalities / new components)
System End of Life / Cessation		EL1	SoI is subject of retract process(es)
		EL2	SoI is subject of dismantling process(es)

Figure 3.7. *Phases and functioning modes checklist [CHA 13a]*

– *Interface functional architecture*: the interface must transform one or more flows stemming from a system source (emitter processor role) to a (set of) system destination(s) (receptor processor role). This transformation allows avoiding physical effects that may impair the systems in interaction (i.e. disturbing or damaging structure/organization or behavior), and moreover must verify the interoperability requirements. We propose modeling the expected transformation: (objective 1) by a model of time, shape and space attribute transformation of the flow and of items transported by the flow and (objectives 1 and 2) by the effect model proposed above focusing on the potential effects to be avoided and anticipated. The functional vision of an interface highlighting these two transformations is then modeled by using eFFBD (enhanced Functional Flow Block Diagram [SEI 09]). Figure 3.8 shows a very simple and abstract depiction of a functional architecture pattern with more specialized patterns presented in Figures 3.9 and 3.10.

– *Interface physical architecture*: the interface is implemented by linking various sub-processors (physical subsystems or components, actors, sub-organizational units), on which the functions proposed in the functional architectures are to be allocated. This description can be generated using, for instance, PBD language (Physical Block Diagram).

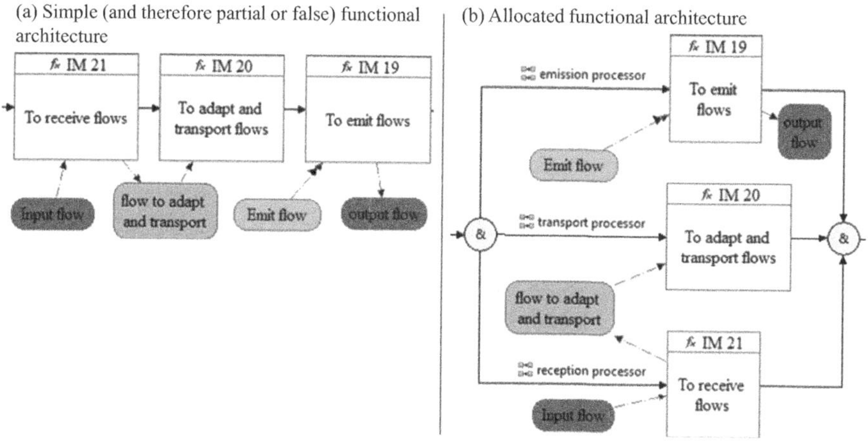

Figure 3.8. *Basic functional architecture patterns: receiving, transporting and sending a flow*

3.5. Conclusion and further work

This chapter has introduced a conceptualization and formalization of requirements that have served to enhance the more classical requirements checklists; this effort has enabled building elements of a pattern model that features a characterization of interface functions, dynamic and physical interface aspects and effects to help designers broaden their vision of the required interactions between the SOI and other systems throughout its life cycle. As a synthesis of results, let us recall various data conceptual models even called meta-models exist in the SE domain. Some of which are even partially standardized and are often implemented in tools, e.g. [COR 14]. The retained concepts (requirement, function, flow, item, link, component, etc.) and relations between concepts (allocatedTo, isTriggerFlowOf, translates, etc.), constraints and required SST attributes, which have to be handled by engineers throughout the SOI design, are then available. The meta-model partially illustrated in the Appendix (section 3.6) summarizes the proposed enhancements to define an interface meta-model for assisting with the nativcly interoperable system design process. This meta-model describes an abstract syntax of the proposed SE modeling language. The resulting modeling tool based on formal framework proposed by [NAS 14] is currently under development and test.

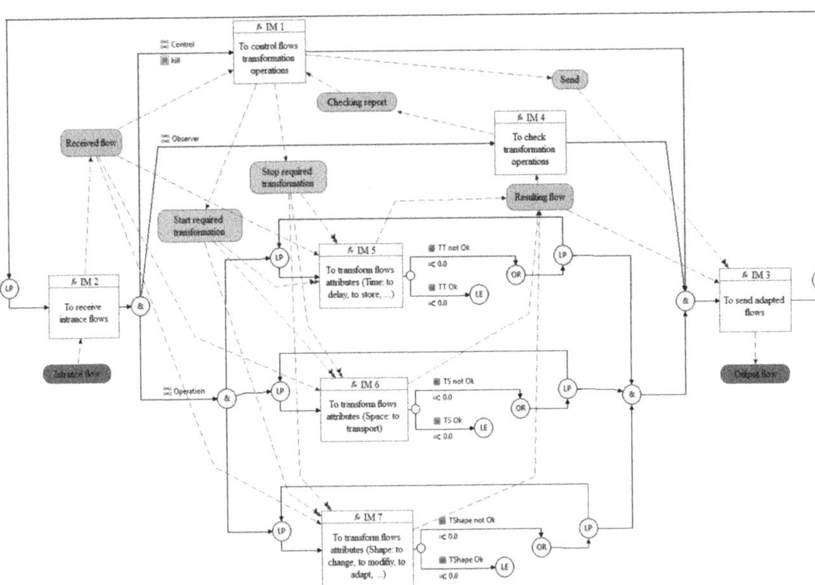

Figure 3.9. *Interface functional pattern: receiving, adapting, transporting and sending a flow*

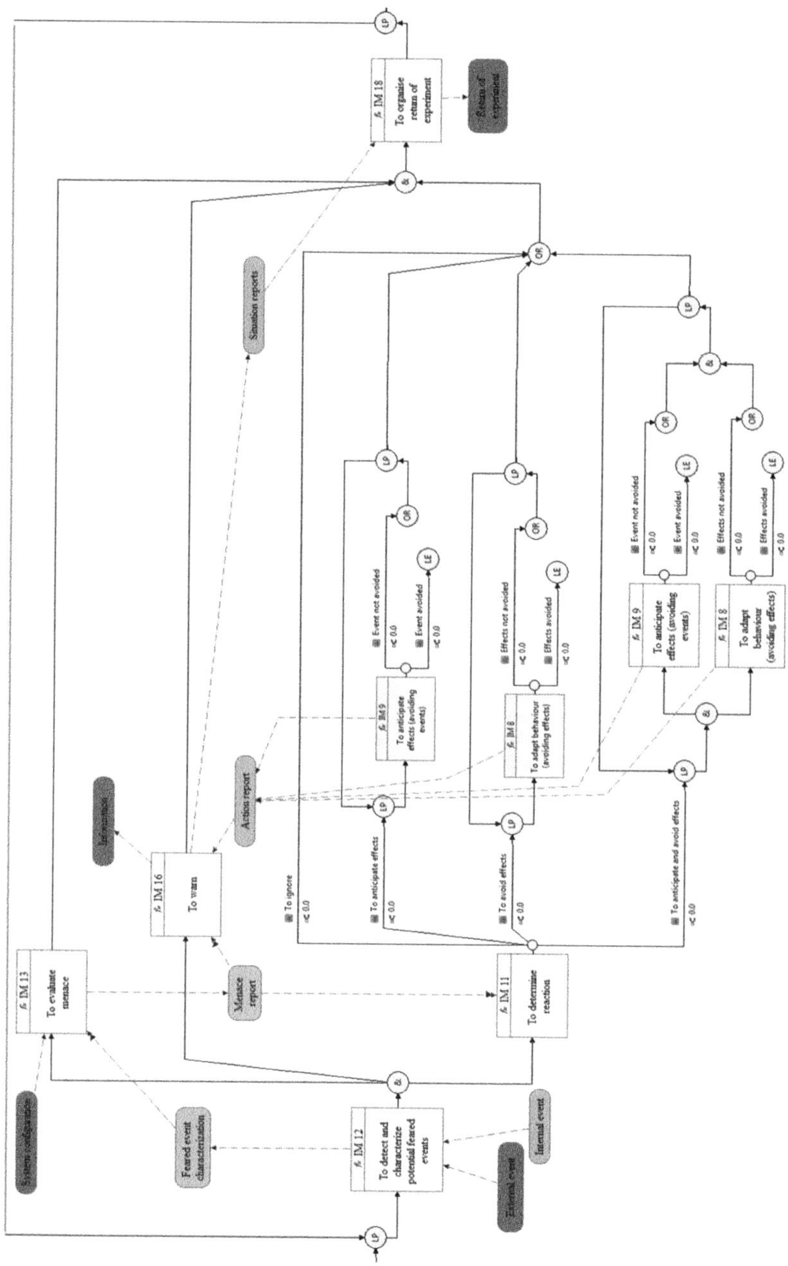

Figure 3.10. *Interface functional pattern: protecting the SOI*

3.6. Appendix

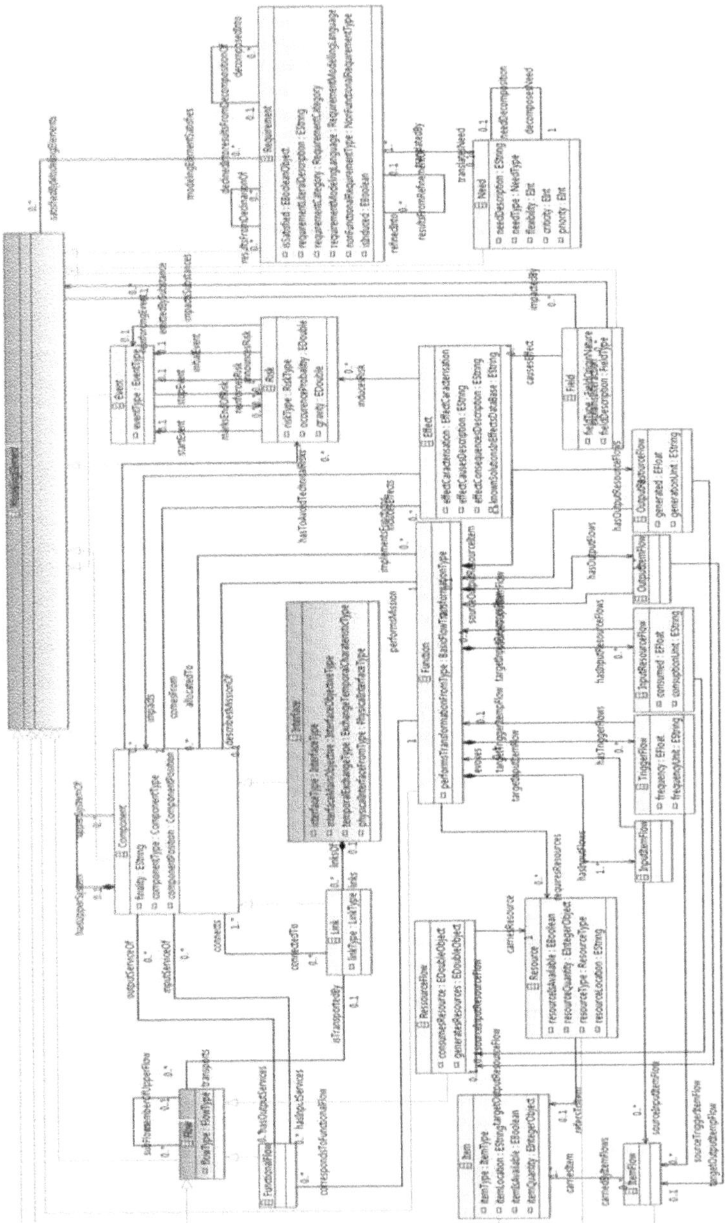

3.7. Bibliography

[AFI 06] AFIS, Interfaces techniques et architectures du système, CT AIS, fiche no. 2, 2006.

[ALT 12] ALTSHULLER G., "The TRIZ method: numerous references and a presentation of TRIZ principles", available at: http://www.altshuller.ru/world/eng/index.asp, 17 July 2012.

[BER 01] BÉRARD B., BIDOIT M., FINKEL A. *et al.*, *Systems and Software Verification: Model Checking Techniques and Tools*, Springer, Berlin, 2001.

[BIL 14] BILAL M., DACLIN N., CHAPURLAT V., "System of Systems Design Verification: Problematic, Trends and Opportunities", in *Enterprise Interoperability VI*, Springer, Berlin, 2014.

[BIL 15] BILLAUD S., DACLIN N., CHAPURLAT V., "Interoperability as a key concept for the control and evolution of the system of systems (SoS)", *6th International Workshop on Enterprise Interoperability, to Appear in IWEI Proceedings*, Nimes, France, May 2015.

[BKC 14] BKCASE EDITORIAL BOARD, The Guide to the Systems Engineering Body of Knowledge (SEBoK), v. 1.3. R.D. Adcock (EIC), The Trustees of the Stevens Institute of Technology, Hoboken, available at: http://www.sebokwiki.org/ (accessed on 17 July 2014), 2014.

[BLA 11] BLANCHARD B.S., FABRICKY W.J., *Systems Engineering and analysis*, 5th ed., Prentice Hall, Upper Saddle River, 2011.

[CAL 07] CALIFORNIA DEPARTMENT OF TRANSPORTATION, System Engineering guide book for transportation systems, Report, Division of Research and Transportation, 2007.

[CHA 13a] CHAPURLAT V., DACLIN N., "Proposition of a guide for investigating, modeling and analyzing system operating modes: OMAG", *International Conference on Complex System Design and Management CSDM 2013*, Paris, December 2013.

[CHA 13b] CHAPURLAT V., "UPSL-SE: a model verification framework for systems engineering", *Computers in Industry*, vol. 64, pp. 581–597, 2013.

[CHA 15] CHAPURLAT V., "Property concept and modelling language for Model-Based Systems Engineering (MBSE) context", *Internal Research Report RR 15-01 (access on demand)*, 2015.

[CLA 99] CLARK T., JONES R., Organisational interoperability maturity model for C2, Report 1999 Command and Control Research Technology Symposium, United States Naval War College, Newport, 1999.

[COR 14] CORE V7.0, "Vitech corporation", available at: http://www.vitechcorp.com/ (accessed on 16 July 2014), 2014.

[CRA 14] CRADLE TOOL, "3SL", available at: http://www.threesl.com/index.php (accessed on 17 July 2014), 2014.

[DAC 08] DACLIN N., CHEN, D., VALLESPIR, B., "Methodology for enterprise interoperability", *Proceedings of the 17th World Congress, The International Federation of Automatic Control*, Seoul, Korea, 6–11 July 2008.

[DAS 10] DASGUPTA P., *A Roadmap for Formal Property Verification*, Springer, Berlin, 2010.

[DOD 01] DoD, "Systems Engineering Fundamentals", Defense Acquisition University Press, available at: http://www.dau.mil/pubscats/PubsCats/SEFGuide2001-01.pdf (accessed on 7 July 2012), 2001.

[DOD 07] DoDAF, Architecture Framework Version 23 April 2007, version 1.5 (Vols I, II and III) (updated version 2.02 is available), 2007.

[EIA 98] EIA/ANS 632, Processes for Engineering a System, Standard, 1998.

[FÉL 07] FÉLIOT C., "Toward a Formal Theory of Systems", *Colloque d'Automne du LIX 2007 – CAL07 Complex Systems: Modelling, Verification and Optimization*, Paris, Carré des Sciences, 3 and 4 October 2007, available at: http://www.lix.polytechnique.fr/~liberti/cal07/presentations/ (accessed on 11 July 2012).

[FER 06] FERRIS T.L.J., "It Depends: Systems of systems engineering requires new methods if you are talking about new kinds of systems of systems", *INCOSE*, 2006.

[FIO 12] FIORÈSE S., MEINADIER J.P., *Découvrir et Comprendre l'Ingénierie Système*, CEPADUES Editions, 2012.

[FOR 11] FORMAL VERIFICATION TOOLS OVERVIEW WEB SITE, available at: http://anna.fi.muni.cz/yahoda/ (accessed on 10 April 2011), 2011.

[GRU 11] GRUHN P., "Human machine interface (HMI) design: the good, the bad, and the ugly (and what makes them so)", *66th Annual Instrumentation Symposium for the Process Industries*, Texas A&M University, 27–29 January 2011.

[HOL 11] HOLLNAGEL E., PARIES J., WOODS D. *et al.*, (eds), *Resilience Engineering in Practice: A Guidebook*, ASHGATE Publishing, Farnham, 2011.

[IEE 05] IEEE, IEEE standard for application and management of the systems engineering process, 1220–2005 Standard, 2005.

[INC 03] INCOSE, System Engineering (SE) Handbook, A Guide For System Life Cycle Processes And Activities Version 3.2.2, INCOSE TP 2003 002 03.2.2, 2011.

[INC 08] INCOSE, Survey of Model-Based Systems Engineering (MBSE) Methodologies, INCOSE-TD-2007-003-01,Version/Revision: B, 10 June 2008.

[INC 14] INCOSE, "SE tools database", available at: http://www.incose.org/ProductsPubs/products/toolsdatabase.aspx (accessed on 7 July 2014), 2014.

[ISO 08] ISO/IEC, IEEE Standards 15288.2008, Systems Engineering – System Life Cycle Processes, February 2008.

[JEA 97] JEAN-MICHEL P., La modélisation par les systèmes en situations complexes, PhD Thesis, Paris Sud University, France, 1997.

[KOP 07] KOPACH-KONRAD R., LAWLEY M., CRISWELL M. *et al.*, "Applying systems engineering principles in improving health care delivery", *Journal of General Internal Medicine*, vol. 22, no. Suppl. 3, pp. 431–437, 2007.

[LAL 97] LALLI V.R., KASTNER R.E., HARTT H.N., Training Manual for Elements of Interface: Definition and Control, NASA Reference Publication RP1370, January 1997.

[LEM 94] LE MOIGNE C., *La théorie du système général: théorie de la modélisation*, PUF, Paris, 1994.

[MAI 98] MAIER, M.W., "Architecting principles for systems-of-systems", *Systems Engineering*, vol. 1, pp. 267–284, 1998.

[MAL 10] MALLEK S., DACLIN N., CHAPURLAT V., "Towards a conceptualisation of interoperability requirements", *International Conference on Interoperability for Enterprise Software & Applications, IESA 2010*, 14–15 April, Coventry, England, 2010.

[MAL 11] MALLEK S., DACLIN N., CHAPURLAT V., "An approach for interoperability requirements specification and verification", *The International IFIP Working Conference on Enterprise Interoperability*, Stockholm, Sweden, 2011.

[MAN 02] MANN D., *Hands on Systemic Innovation*, CREAX Press, 2002.

[MIC 08] MICOUIN, P., "Toward a property based requirements theory: system requirements structured as a semilattice", *System Engineering*, vol. 11, no. 3, pp. 235–245, 2008.

[MOD 10] MODELICA, "A unified object-oriented language for physical systems modeling language specification version 3.2, 24 March 2010", available at: http://www.modelica.org/ (accessed on 16 July 2014).

[MOT 14] MODELICA, "Tools", available at: https://www.modelica.org/tools (accessed on 16 July 2014), 2014.

[NAS 07] NASA, NASA Systems Engineering Hand Book, 1st ed., NASA-SP 2007-6105, December 2007.

[NAS 14] NASTOV B., CHAPURLAT V., DONY C. *et al.*, "A verification approach from MDE applied to model based system engineering: xeFFBD dynamic semantic", *International Conference on Complex System Design and Management CSDM 2014*, Paris, France, December 2014.

[NAU 08] NAUDET Y., LATOUR T., CHEN D., "A systemic approach to interoperability formalization", *Proceedings of the 17th World Congress IFAC, International Federation of Automatic Control*, Seoul, Korea, 6–11 July 2008.

[NUS 00] NUSEIBEH B., EASTERBROOK S., "Requirements engineering: a roadmap", *ICSE 2000 Proceedings, Conference on the Future of Software Engineering*, New York, 2000.

[OXF 12] OXFORD CREATIVITY, "Effects data base", available at: http://wbam2244.dns-systems.net//EDB_Welcome.php (accessed on 27 November 2012), 2012.

[QAM 12] QAMAR A., PAREDIS C.J.J., "Dependency modeling and model management in mechatronics", *Proceedings of ASME 2012, International Design Engineering Technical Conference and Computers and Information Engineering Conference, IDET/CIE 2012*, Chicago, IL, August 2012.

[RGA 05] RGAERO 000 77, Programme Management – Guide for the Management of Systems Engineering, BNAE (Bureau de Normalisation de l'Aéronautique et de l'Espace), 2005.

[SA 14] SA, "IBM rational system architect", available at: http://www-03.ibm.com/software/products/en/ratisystarch (accessed on 16 July 2014).

[SEI 09] SEIDNER C., Vérification des EFFBDs: Model checking en Ingénierie Système, PhD Thesis, Nantes University, 3 November 2009.

[SYS 14a] SYSML, "System Modeling Language Sysml", available at: http://www.sysml.org/ (accessed on 16 July 2014), 2014.

[SYS 14b] SYSML, "Tools", available at: http://www.sysmlforum.com/sysml-tools/ (accessed on 16 July 2014), 2014.

[THI 02] THIMBLEBY H., BLANDFORD A., CAIRNS P. *et al.*, "User interface design as systems design, people and computers XVI- memorable yet invisible", *Proceedings of HCI 2002X.Faulkner*, 2002.

[US 00] US DEPARTMENT OF THE TREASURY CHIEF INFORMATION OFFICER COUNCIL, Treasury Enterprise Architecture Framework. Version 1, July 2000.

[VVT 11] V&V TOOLS, RPG reference document, 2006, available at: http://vva.msco.mil/Ref_Docs/VVTools/vvtools-pr.PDF (accessed on 10 April 2011), 2006.

[WEC 12] DE WECK O.L., ROSS A.M., RHODES D.H., "Investigating relationships and semantic sets amongst system lifecycle properties (Ilities)", *Third International Engineering Systems Symposium CESUN 2012*, Delft University of Technology, 18–20 June 2012.

[WHE 10] WHEATCRAFT L.S., "Everything you wanted to know about interfaces, but were afraid to ask", *INCOE International Symposium*, Chicago, IL, 2010.

[WIK 14] WIKIPEDIA, "-Ilities Reference List", available at: http://en.wikipedia.org/wiki/List_of_system_quality_attributes (accessed on 12 June 2014), 2014.

4

Software Development and Interoperability: A Metric-based Approach

4.1. Introduction

The present research work is aimed to provide the adequate environment for implementing the method proposed in Camara *et al.* [CAM 15] for the validation and verification of interoperability requirements. A metric-based approach for software development and interoperability is proposed:

1) The method for the validation and verification of interoperability requirements is defined in Camara *et al.* [CAM 15] in its generic form. We hypothesize that several choices have to be made when applying this method and every combination of these choices gives a specific definition of the method. The first level of these choices consists of the different axes of the proposed framework (i.e. Process PIs, Analysis and System state evolution).

2) Once the specific definition of the method is obtained, it is possible to evaluate the state of available data and to decide if the specific definition of the method can be implemented or not.

Chapter written by Mamadou Samba CAMARA, Rémy DUPAS and Yves DUCQ.

Section 4.2 presents a review of the literature. The metric-based approach for software development and interoperability is presented in section 4.3. Two illustrative examples are used in the fourth section to demonstrate the applicability of the proposed approach and the final section is dedicated to the conclusion.

4.2. Literature review

4.2.1. *Literature of software requirements' verification and validation*

After the validation activity, the software system is delivered to the customer and is installed and put into practical use. When the validation activity reveals problems, which means that the system is not good enough for use, further development is required to fix the identified problems [SOM 11]. Software system requirements are often classified as functional and non-functional requirements [SOM 11]. Software validation or, more generally, verification and validation is intended to show that a system both conforms to its specification and meets the expectations of the system customer [BOE 79, SOM 11]. Except for small programs, systems can not be tested as a single, monolithic unit. The testing process is made up of three stages: development testing, system testing and acceptance testing, in which system components are tested, then the integrated system is tested, and finally the system is tested on customer's data [SOM 11]. The aim of verification is to check that the software meets its stated functional and non-functional requirements [BOE 79, CHA 08, SOM 11]. The aim of validation is to ensure that the software meets the customer's expectations (i.e. expectations of the organization that commissioned the system) [BOE 79, SOM 11]. Properties expressed as quantitative measures can be naturally verified [BAR 06]. Properties that refer to subjective feeling can be difficult to verify and are a natural target for validation [BAR 06].

4.2.2. *System state evolution*

According to Doumeingts [DOU 01], developing a method dedicated to managing the evolution of an enterprise (or inter-enterprise collaboration) requires the definition of a reference model for evolution. The reference model consists of a sequence of states representing the evolution of the status of the enterprise (or inter-enterprise collaboration). The "As-is" state corresponds to how activities (the physical system) and decisions (the decisional system) are currently performed in the studied inter-enterprise collaboration [CAN 98]. The model of the "As-is" state enables faster diagnostic and more efficient end-to-end process analysis [MED 12]. The "To-be" state represents the expected transformations of the "As-is" models resulting from the implementation of an interoperability solution [CAN 98].

4.2.3. *Interoperability literature review*

A review of the literature was conducted to analyze how the activities of software specification, software development and software validation [SOM 11] were carried out in the interoperability domain.

4.2.3.1. *Interoperability requirements*

Approaches used to represent interoperability requirements in the literature can be summarized with their corresponding limitations as follows:

1) The maturity models [TOL 03, MOR 04, KIN 05]: repetition, ambiguity, imprecision and incoherence because the needs are expressed in natural language [MAL 12].

2) Formal representation of interoperability requirements [MAL 12, COR 12]: the implementation of this approach may experiment scalability problems [CAM 14].

3) Interoperability requirements as problems:

a) requirements are specified by means of collaboration process models: the requirements are disconnected from existing systems and the interoperability problems to solve;

b) the interoperability matrix [CHE 08, DUC 12]: does not provide a structured set of requirements [INC 12].

4.2.3.2. *Collaborative Information System (CIS) architecture*

A CIS aims to support "Information Systems (IS) Interoperability", that is to say, to satisfy requirements such as data conversion, application sharing and process management [TOU 07]. The CIS should then include two different parts: connectors to be plugged into partners' information systems and the concrete entity managing the collaboration: an intermediate information system. Several research works were conducted in order to find logical and technical solutions for the CIS. Approaches proposed in these works can be categorized in two groups: Model-Driven Interoperability (MDI) approaches [CHU 13, KHA 13] and Business Process Lifecycle (BPL) based approaches [FRA 12]. It can be noticed that, in all approaches for CIS development, the proposed platforms are based on Service Oriented Architectures (SOA) [PES 08]. The specification of requirements proposed in the approaches for the development of CIS does not provide sufficient information to describe interoperability problems and then to facilitate the software development activity.

4.2.3.3. *SOA testing*

Since our problem is related to the verification and the validation of interoperability, we will focus primarily on testing approaches and techniques for service-based systems. For the testing of SOA applications, Wieczorek and Stefanescu [WIE 10] identified four distinct testing layers: unit testing, service testing, integration testing and system testing. Lee [LEE 09] and Vieira *et al.* [VIE 06] advocated for the realization of system testing at a process level. The first limitation of this work [LEE 09] is the fact that the business process performance metrics used in the test process are not defined. In our opinion, although interoperability implementation is generally based on SOA, the testing approaches proposed for service-based systems are not suitable for testing inter-enterprise interoperability achievement. Indeed, these approaches [LEE 09, VIE 06, WIE 10] do not reference interoperability problems, which makes them useless for verifying the elimination of the latter.

4.2.4. *The method for the validation and verification of interoperability requirements*

The research work in Camara *et al.* [CAM 15] aims to develop an approach to reach inter-enterprise interoperability and to test its achievement using practices from the software engineering process. Four fundamental activities are identified in the software process: software specification, software development, software validation and software evolution [SOM 11]. For each of the first three activities, our approach proposes defining a sub-goal and determines how to achieve it:

1) *Software specification.* Interoperability requirements specification ensures certain characteristics: requirements are structured (i.e. understandable and identifiable) and testable.

2) *Software development.* The structured nature of interoperability requirements will ease the logical and technical architecture definition.

3) *Software validation.* The testability of interoperability requirements facilitates the definition of a method that allows to test the level of achievement of interoperability improvement. The objective will be to prove that the implementation of the CIS has improved the interoperability up to the desired level.

4.2.4.1. *Improving the interoperability requirements specification*

The interoperability requirements specification proposed in our work results in the definition of two categories of requirements:

1) *Measurable interoperability requirements*. This category concerns measurable interoperability requirements that can be verified in the testing activity. Indeed, according to Sommerville [SOM 11], non-functional requirements should be written quantitatively, whenever possible, so that they can be objectively tested. The measurable interoperability requirements are metrics representing the desired interoperability level for each business process as a result of the implementation of the CIS. These measurable requirements are related to three Performance Indicators (PIs) considered by Camara *et al.* as sufficient to measure interoperability [CAM 14]: the average elapsed time, average cost and percentage of failure.

2) *Non-measurable interoperability requirements*. This category contains interoperability requirements that can be validated in the testing activity but not verified because of their non-measurable nature. For this category, the requirements specification consists of directly representing interoperability problems in business process models, mainly in "As-is" ones. The adopted representation is based on a principle that consists of distinguishing between business activities and Non-Value-Added (NVA) activities, mainly inspired by the work done in Camara *et al.* [CAM 14]. Business activities create value in a business process. NVA activities are defined as the components of business processes that represent efforts between partners to achieve interoperability in information exchange. According to Hammer [HAM 96], Non-Value-Adding work creates no value for customer but is required in order to get the value-adding work done. One of the most important principles of Business Process Reengineering (BPR) is the value-focus principle. The value-focus principle states that NVA activities must be targeted for elimination in order to save time and/or money [HAM 93].

Interoperability problems are depicted in BPMN (Business Process Model and Notation) [OMG 11] business process models. The stereotypes [FRA 03], generally used for the UML language, will be used to differentiate NVA from business activities in the BPMN process models. BPMN is more suitable for modeling collaborative processes because it helps situate the boundaries of the collaborating companies using pools. The proposed representation technique enables us to overcome the limitations presented in section 4.2.3, since the requirements will relate interoperability problems to target elements such as people, organization units and material resources which are clearly identifiable in the collaborating enterprises.

Interoperability requirements are part of non-functional requirements. Both measurable and non-measurable requirements defined in this research work are considered as non-functional requirements because of their relation to interoperability. Indeed, according to Sommerville [SOM 11], non-functional requirements arise through user needs, because of budget constraints, organizational policies or the need for interoperability with other software or hardware systems.

4.2.4.2. *Facilitating architecture definition*

Software development is the activity in which software is designed and programmed. The requirements specification results are inputs of the design and implementation processes [SOM 11]. In order to develop solutions for interoperability problems, the interoperability matrix utilizes the concepts of solution space [CHE 08, DUC 12]. The solution space is composed of the three dimensions of the INTEROP framework. The cross of an interoperability barrier, an interoperability concern and an interoperability approach includes the set of solutions to breakdown the same interoperability barrier for the same concern and using the same approach. In order to correctly determine the solution (third dimension), there must be sufficient information to describe interoperability problems (two first dimensions). The interoperability requirements representation in process models contains the following information: tasks where interoperability problems arise as well as the resources (human and non-human) involved in the interoperability problems. This set of information will facilitate the design process and then improve the software development activity.

4.2.4.3. *Adapting validation activity*

In the Software Engineering (SE) process, the validation activity can be considered as a test process that can be divided into a set of test sub-processes defined to perform a specific test level (e.g. system testing, acceptance testing) or test type (e.g. usability testing, performance testing) within the context of an overall test process for a test project [ISO 13]. A test type is a group of testing activities that are focused on specific quality characteristics [ISO 13]. In the validation activity of a software process, the improvements will consist of considering interoperability as a quality characteristic and developing the interoperability test type. On the basis of the recommendations given in the literature on service-based systems testing, the interoperability testing sub-process will be executed at the system testing level using business process models. The interoperability testing sub-process is divided into two steps: verification and validation of interoperability requirements.

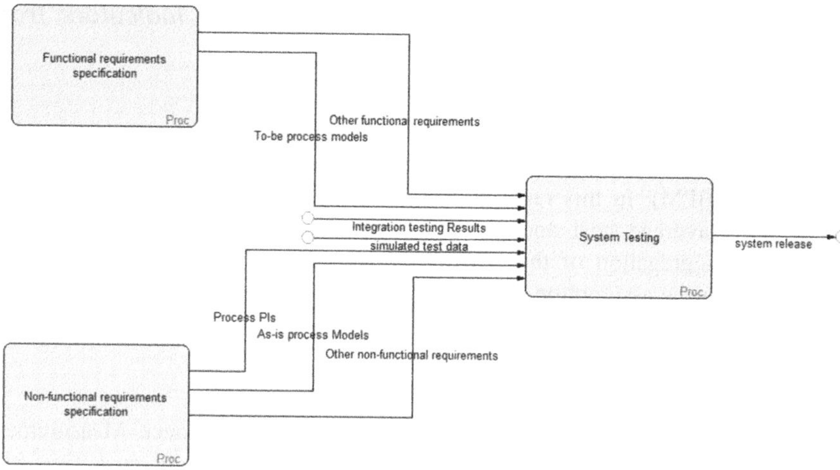

Figure 4.1. *The interoperability testing sub-process*

1) The first step of the interoperability testing sub-process is the verification of measurable interoperability requirements. The goal of this step is to verify if the level of interoperability is improved at the level expected from the implementation of the CIS as determined in the software specification activity. When the percentage of interoperability improvement is judged unsatisfactory then the validation step is launched in order to identify, understand and possibly resolve interoperability problems.

2) The second step of the interoperability testing sub-process is the validation of non-measurable interoperability requirements. The main input of this step (Figure 4.4.1) is the "As-is" business process models which contain the interoperability requirements specification (section 4.2.4.1). The validation is carried out by executing each process in order to verify if all the NVA activities it contains in its "As-is" version are effectively eliminated by the implementation of the CIS. The execution of a business process may reveal the presence of NVA activities. In this situation, the "As-is" business process model gives sufficient information about the interoperability problems related to the NVA activities concerned. This information will be used to fix the interoperability problems.

4.2.5. *Calculation of business process performance indicators from event logs*

4.2.5.1. *Considered process PIs*

Process performance measurement is an important aspect of Business Process Management (BPM). In this research, the considered process PIs are the average elapsed time, average cost and percentage of failure at the process level and represent the aggregation of the PIs for the activities [YAX 10, IBM 14]. This choice relies on the assumption that these characteristics encompass all other types of dynamic properties of business processes [CAM 14].

4.2.5.2. *Mutual impacts between process PIs*

We propose a framework for the definition of a Performance Measurement System (PMS) (Figure 4.2) made up, in its complete version, of cost, time and quality PIs for all the processes impacted by the implementation of the CIS. To have a comprehensive verification process, interoperability measures defined at process level will be used. Indeed, as it is explained in Camara *et al.* [CAM 14], defining measures at activity level will produce too many indicators and increase the PMS complexity. The framework should be understood as a road map where each state represents additional process PIs on which the verification step of the interoperability testing sub-process is applied using approaches defined in process mining. We assume that the PMS is constructed through a series of states, where S_i denotes the set of indicators in the *state*$_i$.

– The difference $S_i - S_{i-1}$ represents additional indicators defined in order to take into account new concerns defined in S_i. Therefore, S_{i-1} is always included in S_i.

– The directed arcs from S_{i-1} to S_i represent the influence of S_{i-1} to S_i–S_{i-1} that is also a part of S_i.

The four states as well as their influence relationships are defined as follows:

1) State1. The first state concerns time PIs for process instances excluding those containing loops. A loop causes a task to be executed multiple times for a given case [VAN 04, VAN 03]. According to Grigori *et al.* [GRI 04], if the maximum number of loops is high, then the number of loops can greatly differ from instance to instance. In other words, the presence of loops will result in variability in the time and other process PIs between different instances of a given process. In our opinion, the origin of this variability is the random nature of reasons why a process or task can fail.

2) State2. In the second state, the new concern is the measurement of loops in business processes. Traditional workflow systems define loops for repeating parts of the process. Tasks or sets of tasks are iterated for multiple reasons:

– they have not yielded an expected result [GUN 09];

– the workflow management system is configured to act automatically (i.e. pass the case back to the last performer) after a long passivity of performers [VAN 07]. Therefore, we can conclude that the time PIs of state1 influence the failure (or loop) PIs added in state2.

3) State3. The process PIs added in state3 are related to time but the difference with the ones defined in state1 is the fact that they also measure the variability of time caused by loops. The group of indicators in state3 are influenced by the indicators defined in state2 because, according to Muehlen and Rosemann [MUE 00], the quality of the process expressed in the number of failures or loops is an indicator for necessary rework. More generally, it is safe to say that quality characteristics of the business process influence time (and cost).

4) State4. The new process PIs included in state4 are related to cost. These measures receive the impact of time and failure.

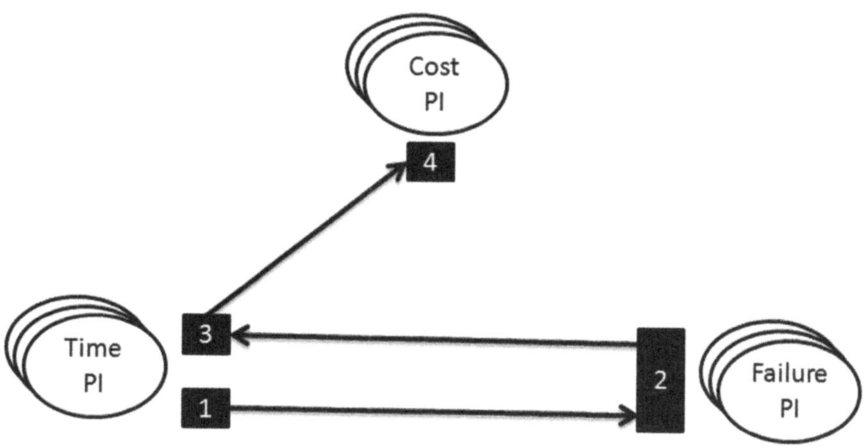

Figure 4.2. *Framework for the definition of the PMS*

4.2.6. *Event logs*

4.2.6.1. *Event logs for internal process*

The Business Process represents a chain of interrelated activities that normally must be connected with the customer requirements [DOU 00]. Process models are built in the design phase and are later used to implement the information system. An "event log" is defined as "a chronological record of computer system activities", which is saved to a file on the system [JAN 13]. Any information system using transactional systems will provide workflow information in some form and to some extent, such as tasks available, their events and the details of these events such as starting/ending timestamp [VAN 03, LIU 12].

4.2.6.2. *Event logs for a collaboration process*

The main data source that can be used for the calculation of process PIs in the verification of interoperability requirements are the integrated event logs which are widely used in the field of cross-organizational process mining. Cross-organizational workflow is usually distributed on different servers owned by different partners or different organizations [VAN 10, ZEN 13]. Cross-organizational workflows are enabled by web-services technology [VAN 07]. A framework for workflow integration based on process mining is proposed in Zeng *et al.* [ZEN 13]. The integrated log results from the integration of the running logs of two or more organizations. In the running log collected from each organization, an event record contains Case(Id), Activity (Name), Start time, End time, Required resources, Released resources, Messages received and Messages sent. The data integrated between the running logs contains an additional column labeled "Organization".

4.2.6.3. *Maturity levels for event logs*

According to Van Der Aalst [VAN 11], there are several criteria to judge the quality of event data:

– events should be trustworthy;

– event logs should be complete, i.e. given a particular scope, no events may be missing;

– any recorded event should have well-defined semantics;

– moreover, the event data should be safe in the sense that privacy and security concerns are addressed when recording the events.

Similarly, Jagadeesh Chandra Bose *et al.* [JAG 13] distinguish four broad categories of problems with regard to the quality of an event log: Missing Data, Incorrect Data, Imprecise Data and Irrelevant Data.

In order to evaluate event logs with regard to the above criteria, [VAN 11] defines five event log maturity levels ranging from excellent quality to poor quality:

1) Highest level: the event log is of excellent quality (i.e. trustworthy and complete) and events are well defined (e.g. semantically annotated logs of BPM systems). Events are recorded in an automatic, systematic, reliable and safe manner. Privacy and security considerations are addressed adequately. Moreover, the events recorded (and all of their attributes) have clear semantics. This implies the existence of one or more ontologies. Events and their attributes point to this ontology.

2) Events are recorded automatically and in a systematic and reliable manner, i.e. logs are trustworthy and complete. Unlike the systems operating at level 3, notions such as process instance (case) and activity are supported in an explicit manner. Example: the event logs of traditional BPM/workflow systems.

3) Events are recorded automatically, but no systematic approach is followed to record events. However, unlike logs at level 2, there is some level of guarantee that the events recorded match reality (i.e. the event log is trustworthy but not necessarily complete). Consider, for example, the events recorded by an Enterprise Resource Planning (ERP) system (e.g. tables in ERP systems, event logs of Customer Relation Management systems, transaction logs of messaging systems, event logs of high-tech systems, etc.). Although events need to be extracted from a variety of tables, the information can be assumed to be correct (e.g. it is safe to assume that a payment recorded by the ERP actually exists and vice versa).

4) Events are recorded automatically, i.e. as a by-product of some information system (e.g. event logs of document and product management systems, error logs of embedded systems, worksheets of service engineers, etc.). Coverage varies, i.e. no systematic approach is followed to decide which events are recorded. Moreover, it is possible to bypass the information system. Hence, events may be missing or not recorded properly.

5) Lowest level: event logs are of poor quality. Recorded events may not correspond to reality and events may be missing. Event logs for which events are recorded by hand typically have such characteristics (e.g. trails left in paper documents routed through the organization ("yellow notes"), paper-based medical records, etc.).

For the management of log in the security domain, *The Complete Guide to Log and Event Management* [CHU 10] proposed a maturity curve that stretches from complete log ignorance, log collection and retention, occasional investigation, to periodic log review and then all the way to near-real-time security monitoring.

4.2.6.4. *Learning prediction models from event logs*

Predictions and recommendations based on models learned using historic information can be used in different phases of the BPM life-cycle [VAN 11]:

1) *Execution phase.* This can be used to guide running process instances. For example, it is possible to predict the remaining processing time of a case. Based on such predictions, we can also build recommender systems that propose particular actions to reduce costs or shorten the flow time.

2) *Adjustment phase.* During the execution phase, the process is monitored. Moreover, smaller adjustments may be made without redesigning the process.

3) *Diagnosis phase.* In the diagnosis phase, the enacted process is analyzed and the output of this phase may trigger a new process redesign phase.

4.3. Metric-based approach for software development and interoperability

The following sections present the two steps of the proposed approach.

4.3.1. *Data collection framework for the validation and verification of interoperability requirements*

The method for the validation and verification of interoperability requirements is defined in Camara *et al.* [CAM 15] in its generic form. We hypothesize that several choices have to be made when applying this method and every combination of these choices gives a specific definition of the method. The first level of these choices consists of the different axes of the framework (i.e. Process PIs, Analysis and System state evolution):

1) Process PIs. This axis consists of the choice of process PIs (i.e. the average elapsed time, average cost and percentage of failure) to include in the application of the method.

2) Analysis. This axis is used to specify what kind of analysis is realized on process PIs.

a) Effects analysis. The first level of analysis is limited to the process PIs assessment in order to evaluate the impact of interoperability.

i) Monitoring of process PIs. All business processes in the scope of the interoperability project must be executed several times. Each execution of a business process corresponds to a process instance and is recorded in an event log. The event log analysis enables the calculation of process PIs that represent measurable interoperability requirements. For example, we can consider that the scope of a specific definition of the method is limited to the average elapsed time process PIs. In this situation, process mining techniques allow us to automatically mine upper bounds for different key performance indicators (like waiting times, execution times, etc.) of a process by taking into account both the timestamps of tasks in a log and the overall structure of the process model given as input [BUS 09, VAN 04, ROZ 09];

ii) Manual calculation. The manual calculation of process PIs is used in cases where historical data (e.g. event logs) are not available for automatic calculation.

When effects analysis is considered, the analysis of causes is limited to identifying the relation between the removal of NVA activities and the variation of process PIs.

b) Causes and effects analysis. The objective is to try to understand the relations between a performance indicator and its root causes. It is necessary to build a Causal Performance Measurement Model (CPMM) capturing the links between two categories of variables:

i) variables representing process PIs to measure the impact of interoperability;

ii) variables representing causes to measure all elements likely to have an impact on interoperability either positively or negatively. The identification of the causes related to interoperability depends on the process PIs under consideration and is based on the review of the literature.

Two cases can arise during the application of the method, each of which corresponds to a way to obtain the CPMM:

a) Learning prediction models. In this case, event log data is available and can be used to create a quantitative model through the execution of a data mining process. Certain aspects have to be determined when applying the data mining process:

i) Learning technique. Supervised learning techniques like multiple linear regression [HEC 95] or Bayesian networks [BOE 03] can be utilized for this purpose;

ii) Unit of analysis. Determine the right unit of analysis (e.g. process, activity, process instance) for a good definition of the cases (i.e. objects or observations of the sample data).

b) Qualitative learning. The authors propose to build the CPMM using the qualitative reasoning technique [HIN 03] to deal with situations where the data required to build a quantitative model are not available.

i) Qualitative reasoning distinguishes between the tasks of model building and qualitative simulation. The model-building task produces a Qualitative Differential Equation (QDE) model consisting of a set of variables and the constraints among them

[QUA 04]. In a QDE model, each variable is described qualitatively with a quantity space. The quantity space is a finite, completely ordered set of landmark values that represent qualitatively important values in the real number line [QUA 94];

ii) The landmark values depend on the fixed unit of analysis.

3) Status of the system. The third axis enables us to specify the status of the system (e.g. "As-is", "To-be" and "Target" states). In the case where multiple states are included in the application of the method, comparison can be made between results obtained in the different states in order to emphasize evolution in interoperability in terms of:

– values of process PIs;

– structure and parameters of CPMM.

4.3.2. *Evaluation and improvement of available data*

The authors propose verifying if the data available in event logs are sufficient or can be improved to support the specified definition of the method. The first three criteria used to evaluate the available data are all choices made to obtain the specific definition of the method using the framework:

1) Process PIs: data must be available for the chosen process PIs;

2) Analysis: two cases have to be taken into account:

a) Effects analysis: data must be available for all variables measuring the impact of interoperability (i.e. process PIs);

b) Causes and effects analysis: data must be available for both variables measuring the impact and the causes of interoperability at the level where the unit of analysis is fixed;

3) Status of the system: it is necessary to verify that data is available for all the states of the system included in the specific definition of the method.

Additional aspects have to be integrated in the evaluation of the available data:

1) The assessment of interoperability will generally require data to be gathered from multiples enterprises. In the context of inter-enterprise collaboration, this information can be obtained using the cross-organizational workflow model and the integrated event log. Applying the framework for workflow integration based on process mining described in section 4.2.6.2 will help us to obtain the cross-organizational workflow model and the integrated event log.

2) Maturity level. Depending on the type of analysis chosen in the specific definition of the method, a certain maturity level will be required for the event logs (see section 4.2.6.3).

4.4. Application

A couple of specific definitions of the method are used as examples to demonstrate the applicability of the proposed approach.

4.4.1. *Example 1*

The first illustrative example used to demonstrate the applicability of the method involves a supply chain in which an interoperability investment is used to improve the quality of the collaboration [CAM 14]. The partners involved in this collaboration are a customer (an e-commerce company), a stockist (a warehouse owner), a customs agent and the customs administration. At the physical system level of the supply chain, the scope of the illustrative example was limited to a goods entry business process. The specific definition of the method presented in this first research work is characterized by the following choices:

1) *Process PIs axis*. Only the average elapsed time process PI is used in this study. Information needed for the computation of the average elapsed time process PI is the probability related to the different paths and the average elapsed time of the activities in the process model.

2) *Analysis axis*. This work includes causes and effects analysis based on QDE which is a qualitative reasoning technique. The analysis is realized at the business process level. It is not exactly the method that is applied because the process PI average elapsed time is used as a leading indicator (i.e. predictive variable). Lagging indicators are key performance indicators (KPIs) measuring the impact of interoperability at all decision levels (operational, tactical and strategic).

3) *Status of the system axis*. A comparison is made in order to show the evolution of interoperability if the interoperability solution is implemented. The values representing the "As-is" situation are calculated by averaging values given by the partners (i.e. companies belonging to the inter-enterprise collaboration). The interoperability solution corresponds to a specific "To-be" process model and that model enables the calculation of the "To-be" process PI using process simulation capabilities of the Adonis business process management tool [GRO 11]. The KPIs for the "To-be" state are obtained by qualitative predictions based on the QDE model.

4.4.2. *Example 2*

The specific definition of the method presented in the second research work is based on the data mining process model called CRISP-DM and it is characterized by the following choices [MAM]:

1) *Process PIs axis*. Quality characteristics (failure rate metrics or loops);

2) *Analysis axis*. The second research work aimed to develop a method for the construction of a model for the prediction of the final execution status of activities. The dependent variable is a categorical variable representing the final execution status of the activity (Complete, Terminated, Loop). The predictors are variables representing elements which the final execution status of the activity. The application of the method was limited to the Data Understanding phase because impact the event logs definition found in the literature does not include most of the elements we have identified as impacting the final execution status of the activity;

3) *Status of the system axis*. This research work is not dedicated to interoperability analysis but is directly applicable to that issue. Only the "As-is" state is considered because the model is used to understand the interoperability problems in the existing system in order to build better solutions.

Based on the evaluation of available data, this work proposes a new definition and implementation of event logs to enable the collection of all characteristics listed in the Data Understanding phase.

4.5. Conclusion

The method for the validation and verification of interoperability requirements is defined in Camara *et al.* [CAM 15] in its generic form. In the first step of the proposed approach, a framework was used to define a specific definition of the method by making several choices regarding the application of the generic method. In the second step, the state of available data was compared to the data needed by the specific definition of the method and thus the decision of implementing the latter can be eventually taken.

4.6. Bibliography

[BAR 06] BARESI L., PEZZE M., "An introduction to software testing", *Electronic Notes in Theoretical Computer Science*, vol. 148, no. 1, pp. 89–111, 2006.

[BOE 79] BOEHM B.W., "Software engineering: R&D trends and defense needs", in WEGNER P. (ed.), *Research Directions in Software Technology*, MIT Press, 1979.

[BOE 03] BOETTCHER S.G., DETHLEFSEN C., "Deal: a package for learning Bayesian networks", *Journal of Statistical Software*, vol. 8, no. 20, pp. 1–40, 2003.

[CAM 14] CAMARA M., DUCQ Y., DUPAS R., "A methodology for the evaluation of interoperability improvements in inter-enterprises collaboration based on causal performance measurement models", *International Journal of Computer Integrated Manufacturing*, vol. 27, no. 2, pp. 103–119, 2014.

[CAM 15] CAMARA M., DUPAS R., DUCQ Y., "Validation and verification of interoperability requirements", *Enterprise Interoperability: Proceedings of the 6th International IFIP Working Conference (IWEI)*, pp. 39-52, Nîmes, France, May 28-29, 2015.

[CAS 09] CASTELLANOS M., ALVES DE MEDEIROS A.K., MENDLING J. *et al.*, "Business Process Intelligence", in CARDOSO J., VAN DER AALST W. (eds), *Handbook of Research on Business Process Modeling*, IGI Global, 2009.

[CAN 98] CANTAMESSA M., PAOLUCC E., "Using organizational analysis and IDEFO for enterprise modelling in SMEs", *International Journal of Computer Integrated Manufacturing*, vol. 11, no. 5, pp. 416–429, 1998.

[CHA 08] CHAPURLAT V., BRAESCH C., "Verification, validation, qualification and certification of enterprise models: statements and opportunities", *Computers in Industry*, vol. 59, no. 7, pp. 711–721, 2008.

[CHE 08] CHEN D., DOUMEINGTS G., VERNADAT F., "Architectures for enterprise integration and interoperability: past, present and future", *Computers in Industry*, vol. 59, no. 7, pp.647–659, 2008.

[CHU 10] CHUVAKIN, A., The Complete Guide to Log and Event Management, White Paper, 2010.

[CHU 13] CHUNGOORA N., YOUNG R.I., GUNENDRAN G. *et al.*, "A model-driven ontology approach for manufacturing system interoperability and knowledge sharing", *Computers in Industry*, vol. 64, no. 4, pp. 392–401, 2013.

[COR 12] CORNU C., CHAPURLAT V., QUIOT J.M. *et al.*, "Customizable interoperability assessment methodology to support technical processes deployment in large companies", *Annual Reviews in Control*, vol. 36, no. 2, pp. 300–308, 2012.

[DOU 00] DOUMEINGTS G., DUCQ Y., VALLESPIR B. *et al.*, "Production management and enterprise modeling", *Computers in Industry*, vol. 42, nos 2–3, pp. 245–263, 2000.

[DOU 01] DOUMEINGTS G., "GEM: GRAI evolution method: a case study", *International Journal of Technology Management*, vol. 22, nos. 1–3, pp. 189–211, 2001.

[DUC 12] DUCQ Y., CHEN D., DOUMEINGTS G., "A contribution of system theory to sustainable enterprise interoperability science base", *Computers in Industry*, vol. 63, no. 8, pp. 844–857, 2012.

[FRA 03] FRANKEL D.S., *Model Driven Architecture: Applying MDA to Enterprise Computing*, Wiley, 2003.

[FRA 12] FRANCO R.D., DE JUAN-MARÍN R., RODRÍGUEZ-MERINO C. *et al.*, "Colnet platform: resilient collaborative networks through interoperability", in POLER R., DOUMEINGTS G., KATZY B. *et al.* (eds), *Enterprise Interoperability V: Shaping Enterprise*, Springer, London, 2012.

[GRI 04] GRIGORI D., CASATI F., CASTELLANOS M. *et al.*, "Business process intelligence", *Computers in Industry*, vol. 53, no. 3, pp. 321–343, 2004.

[GRO 11] GROUP B., Professional BPM with ADONIS NP, https://us.boc-group.com/adonis, September 2011.

[GUN 09] GUNTHER C., Process mining in flexible environments, PhD Thesis, Eindhoven University of Technology, 2009.

[HAM 93] HAMMER M., CHAMPY J., *Reengineering the Corporation: A Manifesto for Business Revolution,* Harper Business, New York, 1993.

[HAM 96] HAMMER M., *Beyond Reengineering*, Harper Business, 1996.

[HEC 95] HECKERMAN D., GEIGER D., CHICKERING D.M., "Learning Bayesian networks: the combination of knowledge and statistical data", *Machine Learning*, vol. 20, pp. 197–243, 1995.

[HIN 03] HINKKANEN A., LANG K., WHINSTON A., "A set-theoretical foundation of qualitative reasoning and its application to the modeling of economics and business management problems", *Information Systems Frontier*, vol. 5, no. 5, pp. 379–399, 2003.

[IBM 14] IBM, Websphere Business Modeler: Process Cost Analysis, IBM, 2014.

[INC 12] INCOSE, Guide for writing requirements, Technical report, INCOSE, 2012.

[ISO 13] ISO/IEC/IEEE, 29119-2-Software and Systems Engineering. Software Testing, 2013

[JAG 13] JAGADEESH CHANDRA BOSE R.P., MANS R.S., VAN DER AALST W.M.P., "Wanna improve process mining results?", *IEEE Symposium on Computational Intelligence and Data Mining (CIDM)*, IEEE, Singapore, pp. 127–134, 2013.

[JAN 13] JANS M., ALLES M., VASARHELYI M., "The case for process mining in auditing: sources of value added and areas of application", *International Journal of Accounting Information Systems*, vol. 14, no. 1, pp. 1–20, 2013.

[KHA 13] KHADKA R., SAPKOTA B., PIRES L.F. *et al.*, "Model-driven approach to enterprise interoperability at the technical service level", *Computers in Industry*, vol. 64, no. 8, pp. 951–965, 2013.

[KIN 05] KINGSTON G., FEWELL S., RICHER W., "An organizational interoperability agility model", Technical report, DSTO Fern Hill, Department of Defence, Canberra ACT 2600, Australia, 2005.

[KUI 94] KUIPERS B., *Qualitative reasoning: modeling and simulation with incomplete knowledge*, MIT Press, Cambridge, 1994.

[KUI 04] KUIPERS B., "Qualitative Simulation" in MEYERS R.A. (ed.), *Encyclopedia of Physical Science and Technology*, 3rd ed., Elsevier, 2004.

[LEE 09] LEE Y., "Event-driven SOA test framework based on BPA-simulation", *First International Conference on Networked Digital Technologies (NDT 2009)*, IEEE, pp. 189–194, 2009.

[LIU 12] LIU Y., ZHANG H., LI C. *et al.*, "Workflow simulation for operational decision support using event graph through process mining", *IDecision Support Systems in Industry*, vol. 52, no. 3, pp. 685–697, 2012.

[MAL 12] MALLEK S., DACLIN N., CHAPURLAT V., "The application of interoperability requirement specification and verification to collaborative processes in industry", *Computers in Industry*, vol. 63, no. 7, pp. 643–658, 2012.

[MAM 15] MAMADOU S., CAMARA F., IBRAHIMA M.G. *et al.*, "Activity failure prediction based on process mining", *the 19th International Conference on System Theory, Control and Computing, ICSTCC – Joint Conference SINTES 19, SACCS 15, SIMSIS 19*, IEEE, pp. 854–859, 2015.

[MED 12] MEDINI K., BOUREY J.P., "SCOR-based enterprise architecture methodology", *International Journal of Computer Integrated Manufacturing*, vol. 25, no. 7, pp. 594–607, 2012.

[MOR 04] MORRIS E., LEVINE L., MEYERS C. *et al.*, "System of systems interoperability (SOSI): final report", Technical report, Carnegie Mellon University, 2004.

[MUE 00] MUEHLEN M.Z., ROSEMANN M., "Workflow-based process monitoring and controlling-technical and organizational issues", *Proceedings of the 33rd Annual Hawaii International Conference on System Sciences*, IEEE Xplore Digital Library, pp. 1–10, 2000.

[OMG 11] OMG, Business Process Model and Notation (BPMN). version 2.0, OMG, 2011.

[PES 08] PESSOA R.M., SILVA E., SINDEREN M.V. *et al.*, "Enterprise interoperability with SOA: a survey of service composition approaches", *Enterprise Distributed Object Computing Conference Workshops*, IEEE, 2008.

[ROZ 09] ROZINAT A., MANS R., SONG M. *et al.*, "Discovering simulation models", *Information Systems*, vol. 34, no. 3, pp. 305–327, 2009.

[SOM 11] SOMMERVILLE I., *Software Engineering*, Addison-Wesley, Boston, 2011.

[TOL 03] TOLK A., MUGUIRA J., "The levels of conceptual interoperability model", *Proceedings of 2003 Fall Simulation Interoperability Workshop*, pp. 1–10, Orlando, FL, 14-19 September, 2003.

[TOU 07] TOUZI J., LORRÉ J.P., BÉNABEN F. *et al.*, "Interoperability through model-based generation: the case of the collaborative information system (CIS)", in GONCALVES R.J., MULLER J.P., MERTINS K. *et al.* (eds), *Enterprise Interoperability. New Challenges and Approaches*, Springer, London, 2007.

[VAN 03] VAN DER AALST W., DONGEN B.F.V., HERBST J. *et al.*, "Workflow mining: a survey of issues and approaches", *Data & Knowledge Engineering*, vol. 47, no. 2, pp. 237–267, 2003.

[VAN 04] VAN DER AALST W., WEIJTERS A., "Process mining: a research agenda", *Computers in Industry*, vol. 53, no. 3, pp. 231–244, 2004.

[VAN 07] VAN DER AALST W., REIJERS H.A., WEIJTERS A. *et al.*, "Business process mining: an industrial application", *Information Systems*, vol. 32, no. 5, pp. 713–732, 2007.

[VAN 10] VAN DER AALST W., "Configurable services in the cloud: supporting variability while enabling cross-organizational process mining", *OTM'10 Proceedings of the 2010 International Conference on the Move to Meaningful Internet Systems*, pp. 8–25, 2010.

[VAN 11] VAN DER AALST W., *Process Mining Manifesto*, Springer, Berlin, 2011

[VIE 06] VIEIRA M., LEDUC J., HASLING B. *et al.*, "Automation of gui testing using a model-driven approach", *The 2006 International Workshop on Automation of Software Test, ACM*, pp. 9–14, 2006.

[WIE 10] WIECZOREK S., STEFANESCU A., "Improving testing of enterprise systems by model-based testing on graphical user interfaces", *17th IEEE International Conference and Workshops on Engineering of Computer-Based Systems*, IEEE, pp. 352–357, 2010.

[YAX 10] YAXIONG T., ZHEN X., HUIBIN X., "BPM exception monitoring based on process knowledge", *Proceedings of IEEE Conference on Cybernetics and Intelligent Systems (CIS)*, IEEE Xplore Digital Library, Singapore, 2010.

[ZEN 13] ZENG Q., SUN S.X., DUAN H. *et al.*, "Cross-organizational collaborative workflow mining from a multi-source log", *Decision Support Systems*, vol. 54, no. 3, pp. 1280–1301, 2013.

5

Decisional Interoperability

5.1. Introduction

Numerous works related to interoperability have shown that, beyond the classical problem of interoperability in ICT, the development of interoperability covers other aspects such as processes, organizations, etc. [ISO 11]. Among these aspects, one is related to the decision-making also known as *decisional interoperability* [DAC 07]. The development of decisional interoperability takes place in the context of decision-making between several partners.

Although this partnership aspect is little considered, decision-making involving several partners is of great importance. Indeed, each decision made by a partner has an impact on other partners and vice versa. In this context, the development of decisional interoperability has to allow easier and better communication and interaction between decision makers. Thus, the objective is not to develop the concepts of coordination and consistency (or else algorithm and method) as in traditional decision-making activity, but the concepts allowing decision makers – who possess their own interest and limitations – to build a common decision together, while remaining independent. As a result, decisional interoperability has to offer the possibility for stakeholders (1) to share fundamental information that is required to build a decision and (2) to use exchanged information to effectively make a decision that respects both the interest of the partnership and the own interest of partners. Hence, it is necessary to clearly highlight and define which kind of information is involved in decisional interoperability and how this information can be used to better organize the decision-making.

Chapter written by Nicolas DACLIN, David CHEN and Bruno VALLESPIR.

The purpose of this part is to present the basic concepts and principles of decisional interoperability. As from now, it is important to note that this chapter does not take the socio-psychological aspect of decision-making into account and focuses on formal and quantitative aspects in order to reduce the uncertainty, in terms of the objectives that the partners have to reach, during a collaborative decision-making process. Furthermore, it also aims to extend the GRAI decisional model concepts to the development of interoperability in the context of networked enterprises.

First, the definition and a mathematical formalization related to the decision-making activity within one enterprise are presented. Set theory is used to formalize the decision-making activity. Then, the decision-making activity in a collaborative context is discussed. Based on these works, a definition of decisional interoperability is proposed. Some design principles to build a decisional interoperability solution are presented and described. Set theory notations are also used to formalize the basic concepts of the decisional interoperability. Running examples are given in each part to illustrate and show the applicability of the proposed approach.

5.2. Decision-making

5.2.1. *Definition*

5.2.1.1. *Decision-making is an activity that aims at making a choice*

The decision itself is the result of a selection of an alternative offered to a decision maker [CEN 04]. In other words, a decision maker has to choose from among a set of alternatives the one that will best allow reaching a defined objective. Among existing models that take an interest in the decision-making activity, Simon's model [SIM 60] is probably the most well-known model. Furthermore, beyond the organization of the decision-making activity through a process, the decision-making activity is also the subject of the development of methods and tools to help with facilitation (e.g. multi-criteria analysis, multi-objective programming, etc.) in terms of automation and to support decision makers in order to make their decision more reliable and efficient [DHA 05, HAM 03]. It is also developed in terms of behavior, understanding, decisions and actions of decision makers. Indeed, although decisional activity tends to be improved – in term of time, search of optimal solutions, etc. – the fact remains that decisional activity is, first of all, a cognitive activity in which it can be difficult to rely on tools – and by extension to substitute human expertise to machine – especially when decision makers face complex situations. In this sense, we can mention contributions such as [RAS 81] that identifies three kinds of behaviors with their associated human mistakes, [END 95] that takes an interest in situational awareness and its steps (perception, comprehension and projection) or else [IMM 14] who considers the representation, the merging and the aggregating of experts' assessments, especially in a critical context (e.g. crisis management).

Thus, to make a decision, a decision maker can follow different clearly identified steps as for instance in the Simon's model (Figure 5.1). This one is made of three main activities know as Intelligence, Design and Choice (IDC).

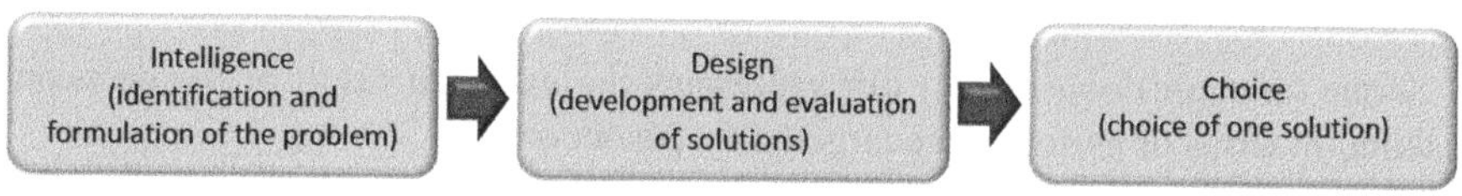

Figure 5.1. *The decision-making process (Simon's model) [SIM 60]*

5.2.1.2. *Identification of the problem (Intelligence)*

Identification corresponds to the formulation of the problem that requires a decision, i.e. a choice among several possibilities. For instance, a given enterprise has to face the problem of including extra order in the production plan.

5.2.1.3. *Development of solutions (Design)*

At this stage, the decision maker looks for solutions that allow him to solve the problem. To this end, the decision maker can use actions means (decision variables) that are available. For instance, for the problem described above, the decision variable can be the use of *overtime* and *subcontracting*. Furthermore, these decision variables are limited in their use (i.e. constrained). Thus, the limited decision variable can be expressed as "*100 h of overtime*" and "*500 h of subcontracting*".

5.2.1.4. *Choice of solution (Choice)*

During this step, the decision maker selects a solution allowing us to meet, at best, the original objective and to solve the problem. For instance, the original objective to reach, for the enterprise, is "*to respect a delivery date*". To this end, the solution is selected among the set of solutions defined in the previous activity. Thus, the decision maker can use *n* hours of subcontracting and *n* hours of overtime in such a way that the choice allows: (1) reaching the objective, i.e. to respect the delivery date, and (2) respecting constraints. If the decision maker is unable to find a correct solution, he can iterate on previous phases in order to correct or to adjust these ones.

5.2.2. *Decision-making in the GRAI model*

Among available decisional tools and models, the GRAI decisional model allows representing a system from a decisional point of view. It makes available a set of concepts to model, analyze and improve the decision layer of an enterprise. In the GRAI model, to make a choice among possible solutions, the decision maker has a set of information (internal or external to the enterprise) that represents a decision frame and influences him [DOU 98, VAL 02]. This decision frame guides the decision maker in its decision-making and is made of five items such as:

– one (or more) *objective*(s) that represents the expected result of the decision-making (in terms of performance targeted by the decision);

– one (or more) *decision variable*(s) that represents the means offered to the decision maker in order to reach the defined objective(s);

– *constraints* that limit the use of decision variable;

– sometimes, *criteria* that guide the decision maker in its choice on the decision variable (e.g. to minimize the use of subcontracting);

– *performance indicators* that represent the results of previous decision-making.

Although this set of concepts is important, the last two concepts are more in support of decision-making while the first three concepts are fundamental to decision-making and their association makes up an area of freedom that constrains and guides a decision maker throughout the decision process. As a result, in the GRAI model, the decision-making can be seen as the "*search for a solution – that meets, at best, the objectives – within a space defined by the decision variables and limited by constraints*" [VAL, 03]. In other words, the decision is a selection of a set of values taken on decision variables that are offered to the decision maker. This set will have to be translated into a performance in order to be compared with the initial assigned objective. Figure 5.2 illustrates a simplified representation of the decision frame and the area of freedom for decision-making.

Thus, to draw a parallel with the previous example, the decision maker has to respect a delivery date. In this sense, overtime and subcontracting can be used (decision variables) and the limits of the use of these decision variables (constraints are 100 h for overtime and 500 h for subcontracting. The area of freedom resulting from this information is given in the following figure. The decision maker has to make a decision within this area in order to reach its assigned objective.

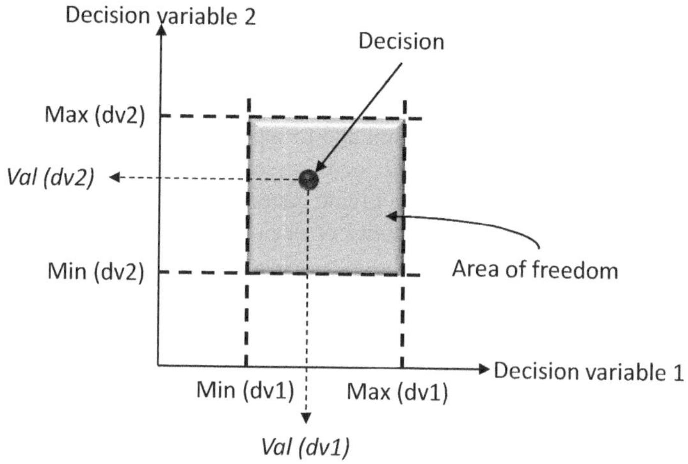

Figure 5.2. *Representation of the area of freedom for decision-making (two decision variables are considered here)*

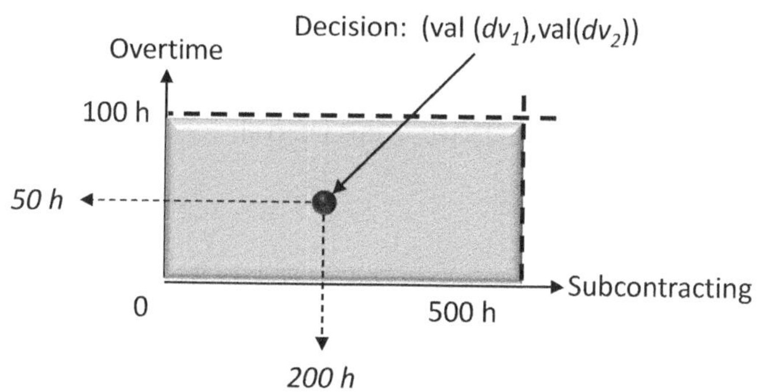

Figure 5.3. *Building of an area of freedom according to decision variables and associated constraints*

In this example, the chosen solution of 50 h of overtime and 200 h of subcontracting is satisfying since it allows reaching the objective and respecting fixed constraints.

5.2.3. *Formal characterization of decision-making in the GRAI model*

Starting from the concepts of the GRAI model previously presented, it is possible to propose a formalization of the decision-making activity. This formalization is necessary to understand and to represent in a precise way the basic concept related to decision-making activity as seen in the GRAI model. Furthermore, this formalization allows giving general rules to make a decision and it can be reused and adapted to other modes of decision-making, including decisional interoperability.

In order to properly characterize the decision-making, the notation coming from the set theory [WEI 08] is used. This one allows expressing clearly and precisely the decision as seen in the GRAI model.

Let OB be the N objectives (ob_1, …, ob_N) that the decision has to ideally achieve. Each objective ob_i takes a performance value on a performance domain P_i.

Let DV be the M decision variables (dv_1, …, dv_M) that can be used by the decision maker to reach the N defined objectives. For each dv_j, a constraint domain ct_j is associated with and corresponds to the values authorized for dv_j. For instance, let us consider the example described in the previous section where the decision variable dv_1 corresponds to the use of overtime. The use of overtime is limited to 100 h so it is possible to set the constraint ct_1 as an interval, i.e. $ct_1 = [0;100]$. Thus, the value of overtime has to be selected in this interval.

Note that a constraint[1] can be:

– an interval: [min, max];

– a union of intervals: [min, max] $\cup$ [min, max];

– an interval including forbidden values: [min, max]\B, where B $\subset$ [min, max] are the forbidden values.

From this consideration, a decision maker can build its area of freedom with M dimensions whose axes are dv_j, which are limited by the constraints ct_j (Figure 5.4). In this example, all constraints are intervals.

1 Even if used values are discrete, we choose the following notation: {1, 2,…, n} = [1, n].

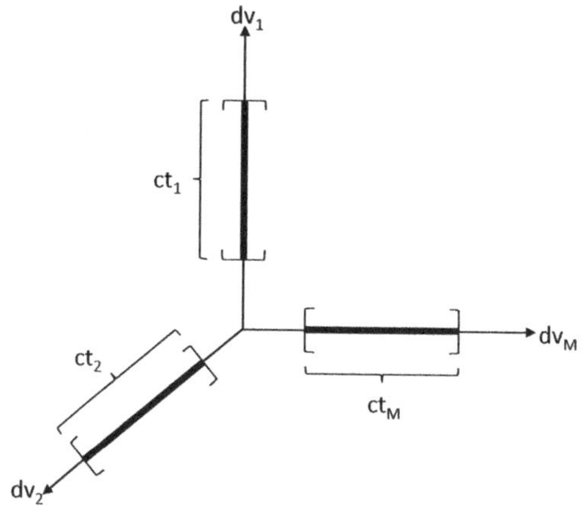

Figure 5.4. *Representation of the area of freedom (M = 3)*

Thus, a decision D corresponds to a vector made of values selected on each decision variable and which respect their own constraints (i.e. the area of freedom available for the decision maker) (Figure 5.5).

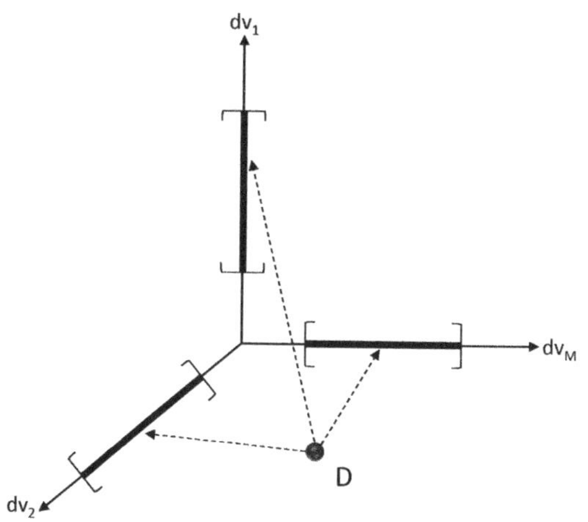

Figure 5.5. *Representation of the vector of decision*

The decision maker has to find a vector that allows reaching the assigned objectives. To achieve this, it may have an optimization tool that provides either the optimal solution or a set of compromise solutions. These solutions are fully adapted and can be implemented in the GRAI model to support decision-making. However, it is important to note that these tools are not always available or understood and their use can be time consuming (in some situations, decision-making has to be fast to address the problem). Therefore, the decision maker relies solely on its skills and its area of freedom to make a decision that respects the objectives. In this sense, he will search for a satisfying solution rather than an optimum.

Let CT denote the set of decision, i.e. $CT = ct_1 \times ct_2 \times \ldots \times ct_m$. For each objective ob_i, let us define a function f_i that expresses the performance obtained on this objective from a given decision, f_i: CT $\rightarrow P_i$.

Thus, the solutions, i.e. performances that can be reached, must be compared to the objectives and, the decision maker will have to choose a solution that meets, at best, the assigned objectives. Some works go further and give to the decision makers the possibility of expressing their preferences on the objectives by using mathematical operators [MON 15].

Figure 5.6 shows the correspondence between the formalization of the concepts of the GRAI decision model and the example of decision-making described in the previous section.

	Formalisation	Application
Objective	ob_i	ob_1: Respect a production capacity (250 hours)
Decision variables	(dv_1, dv_2)	dv_1: overtime (ov) dv_2: subcontracting (su)
Constraints	ct_1 associated to dv_1 ct_2 associated to dv_2	$ct_1 = [0; 100]$ $ct_2 = [0; 500]$
Vector of decisions	$D = \begin{bmatrix} dv_1 \\ dv_2 \end{bmatrix}$	$D = \begin{bmatrix} 50 \\ 200 \end{bmatrix}$
Function	f: sum of each value of the vector	$50+200 = 250$h
Solution	Comparison between acceptable solution and objective	

Figure 5.6. *Formalization of the concept of GRAI decisional model and application*

This part showed the concepts of the decision-making and, especially, those developed in the GRAI decisional model. A formal characterization of decision-making is proposed to show (1) the constituents that participate in the decision-making process and (2) their succession and use to make a decision. This set of concepts and formalization acts as a basis for the development of the interoperability in the decision-making process, namely the decisional interoperability.

5.3. Decisional interoperability

In the context of collaboration, partners work together in order to reach a common goal. At the level of decision-making, it involves the decision makers taking a decision in order to reach the objective of the collaboration and also satisfy their own interests, as well as the process related to this kind of decision being clearly identified (Figure 5.7) [KVA 00].

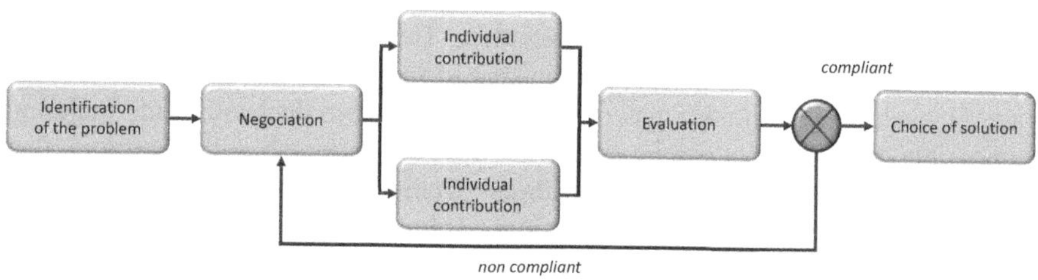

Figure 5.7. *Representation of the collaborative decision-making process*

The main problems lie in the multitude of information coming from different decision makers – that have to be shared to reach a decision – as well as the potential impact of a decision made by a given decision maker on other ones. Thus, decisional interoperability can be defined, initially, as "the ability for different partners to share information that supports decision-making and to use them to make a common decision" [DAC 06]. Hence, the development of decisional interoperability deals with decision-making between several decision makers especially in term of identification, definition and formalization of information to exchange. It has to guide the decision maker during the process in order to address the different problems which handicap the decision process, such as [CHE 05]:

– the majority of decisions among several decision makers is subjective (there is no formalization and no rational manner to take a decision);

– a decision made by a given decision maker impacts the other ones. The decision-making does not always consider the interests of all stakeholders. We find here, the classical systems of decision-making such as: (1) the dictatorship (a decision maker imposes its own decision on others); (2) the consensus (each decision maker gives its consent) or else (3) the vote (each decision maker gives its preferences) [BEN 12];

– a decision is made without analyzing, consistently, all alternatives offered to decision makers;

– numerous iterations are needed to find a solution that satisfies all decision makers.

Therefore, decisional interoperability has to support the collaborative decision-making activity and to prevent the above mentioned problems. Let us consider two cases as shown in Figure 5.8. In the first case (case 1), the decision maker 1 made a decision without knowing either the decision frame or the possible decisions that can be made by the decision maker 2. In this case, decision maker 1 can make a decision that, on the one hand, does not satisfy decision maker 2 and, on the other hand, does not satisfy the objective of the collaboration. This leads to some iterations and negotiations between the two decision makers to find a solution. In the second case (case 2), each decision maker has clearly defined its own decision frame including the possible decisions that can be engaged within the partnership. Furthermore, they share their decision frames and analyze them together to find a solution. In this case, they avoid wasting time and they can find a solution that satisfies both their own interests and those of the partnership.

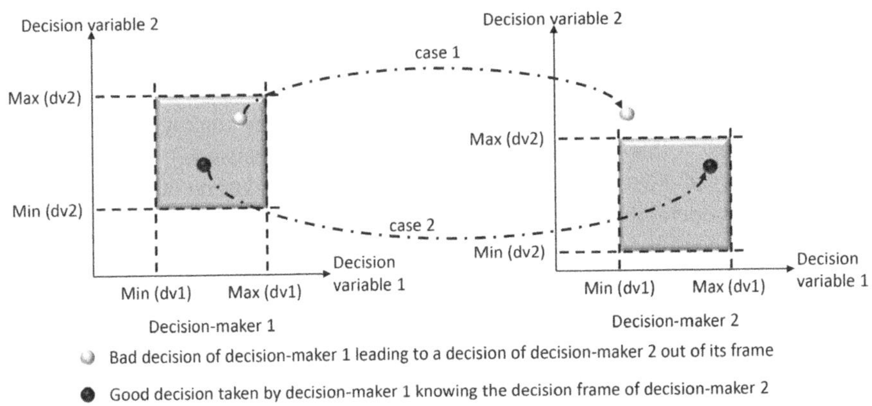

Figure 5.8. *Representation of the problems of collaborative decision-making*

Although the existing collaborative decision-making process is well identified, it is not designed to perform the decisional interoperability that means it does not take a particular interest in the information used and shared within it. Consequently, it is important to define principles for designing decisional interoperability solutions.

5.3.1. *Basic concepts*

Development of decisional interoperability targets two objectives. First, it has to meet its defined requirements (sharing of decision frames that are specific to each decision maker, respect of all decision frames). Within an integrated system, the objectives of each decision level result from the global objective. Thus, a decision frame, at a given level, is a version of a decision frame coming from a higher level. The definition of decision frames is therefore based on a top-down approach. The decision frames from the highest levels are first defined, and then lower ones are defined and so on (Figure 5.9). The GRAI model is based on this principle that allows ensuring consistency throughout the decisional structure.

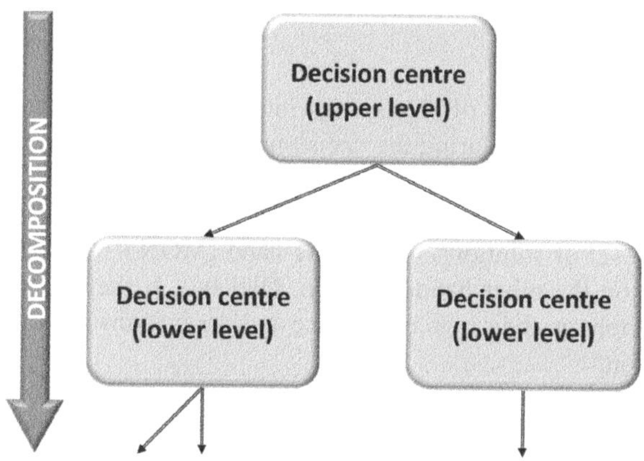

Figure 5.9. *Principle of the decisional decomposition*

However, the interest of such structure is not necessarily that a decision frame – at a given level – results from a decision frame at a higher level. More precisely, it is about ensuring that decision frames belonging to two adjacent decision levels are consistent, i.e. they contribute to the global functioning of the structure. Hence, it is possible to consider a bottom-up approach. Indeed, if two decision centers at a given decisional level are considered and a "virtual" decision center – that is consistent with the two first ones – can arise, then, it means that both decision centers effectively contribute to the global functioning and are controlled by this virtual center. Conversely, it means that this global functioning does not exist. In the context of decisional interoperability, it amounts to considering two independent systems' decisions whose control relies on two decision centers which are also independent. As a result, if two systems work together and they are able to reach a common area of decision that is consistent with these two decision centers, then they are interoperable since they contribute to the global functioning of the partnership (Figure 5.10).

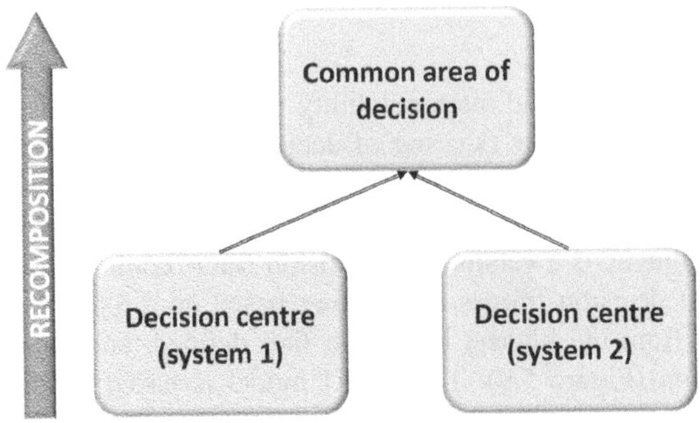

Figure 5.10. *Principle of the decisional interoperability*

Then, the development of decisional interoperability has to overcome the problems that partners face during the decision process. Currently, tools to facilitate using decision-making and the research of solutions among multiple partners exist. Once more, decision makers have the possibility of using tools which allow them to find and make a set of solutions, on the one hand [MON 03], and guiding them to take a solution, on the other hand [PÉN 06, QUI 04]. If they do not possess such tools, they must rely on their own experience and the information they have in order to make a decision.

Unlike the previous works and, in agreement with the GRAI model, the approach proposed here to develop decisional interoperability does not directly address the research of solutions. Therefore, the principle is to show in what manner a common decision space can be built between decision makers, from their own decision frame. This decision frame has to restrict – in an effective way – the area of freedom (i.e. the space in which a decision can be made respecting constraints) of each partner in order to make a decision that respects (1) their own interest(s) and (2) the interest(s) of the partnership.

To continue the development of the decisional interoperability and, thus, to implement the principle defined earlier, design principles for decisional interoperability are formulated. These principles highlight the elements of decisional interoperability and their use to build a decision between several decision makers.

5.3.2. *Design principles for decisional interoperability*

The main purpose of the design principles for decisional interoperability is to provide stringent rules that decision makers can apply in order to make a common decision. Starting from the problems encountered in collaborative decision-making, four fundamental design principles are identified, formulated and described as follows:

– Axiom 1: "*when decision makers intend to ensure decisional interoperability, they must express their own decisional capability.*"

Decisional capability represents the decision potential of a decision maker that means the acceptable limits of each decision maker, regarding the decision requested by the partnership. This capability allows limiting the freedom of decision while remaining within their acceptable limits and has to be explicitly defined.

– Axiom 2: "*when decision makers intend to ensure decisional interoperability, they must compose solutions based solely on their decisional capability.*"

Feasible solutions, in the partnership, have to be made only from the decisional capabilities of each decision maker. Beyond the search of a solution that meets the objectives of the partnership, this avoids a decision maker making a decision that does not comply with the limits of another one.

– Axiom 3: "*when decision makers want to ensure their interoperability, they must build a Common Area of Decision.*"

The construction of a common area of decision, solely on the basis of decisional capabilities of each decision maker and feasible solutions with respect to objectives of the partnership, enables them to make a common decision that respects the objectives of the partnership and their owns limits.

– Axiom 4: "*when decision makers intend to ensure decisional interoperability, they have to implement mechanism to share their decision frame and decisional capabilities.*"

Building a decision frame and decisional capability known by all collaborative decision makers improves decisional interoperability in terms of iterations between parties and the delay before reaching a decision.

The full application of the four axioms for decisional interoperability has to improve it and guide decision makers to build a decision throughout the decision-making process. Finally, the definition of decisional interoperability and its first three associated design principles are formally characterized to clearly define decision-making involving different actors.

5.3.3. *Formal characterization of decisional interoperability*

Based on the formal characterization of the concept coming from the GRAI model (shown in section 5.2.3), the definition of decisional interoperability, its concept as well as its design principles, the formal characterization of the decisional interoperability is proposed below. Following the same approach as for the first formalization, a set of notations that represents interoperability is given. Let us consider the following notation:

– Let *OB'* be the N objectives (ob_1, …, ob_N) that partners have to ideally achieve. Each objective ob_i takes a performance value on a performance domain P_i.

– Let *D* be the M decisions (d_1,…,d_M) that can be taken by a given partner, within the partnership.

Each partner has to set up the set of decision that can be made under the partnership. This set represents the decisional capacity of a decision maker that means its limits in which he can act regarding to its decision frame (Figure 5.11).

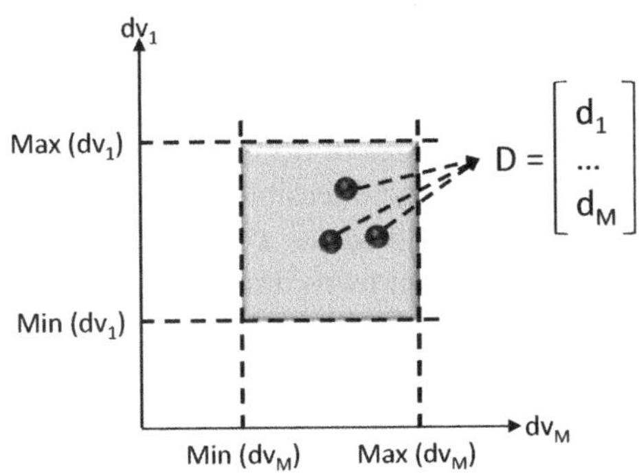

Figure 5.11. *Extraction of the decisional capability of a decision maker*

Drawing up the decisional capacity allows decision makers to make a decision that meets their own interest during collaborative decision-making.

Let R be the set of feasible decisions that can be taken considering all eligible decisions for each partner. Whatever the feasible decision within *R*, this one is characterized by a vector made of values coming from, only, the set of decisions P that decision makers can make within the partnership (Figure 5.12).

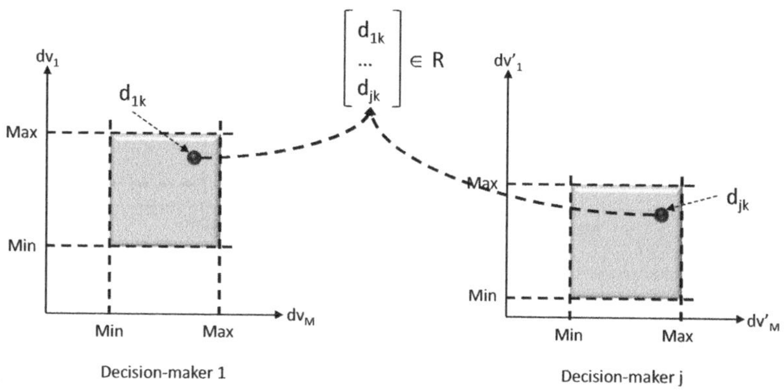

Figure 5.12. *Representation of a vector of feasible solutions and its components*

Let $r \in R, f_i(r)$ be a value that represents a performance p that can be reached for a given objective ob_i ($1 \leq i \leq N$) of the partnership. The feasible solutions, i.e. the performance that can be reached by the feasible decision, have to be compared with the objectives of the partnership. Only the feasible solutions that allow meeting the objectives of the partnership have to be selected to draw up the common decision area of decision.

Let *CAD* be the common area of decision space of the partners. This space is only composed of admissible solutions belonging to the set of feasible solutions. It is in this common area of decision that partners will select a solution allowing them to meet, at best, the objectives of the partnership and their own interests. Once more, decision makers have the possibility to use tools, allowing them to select the best solutions in agreement with common and individual interest. Thus, decision makers have the opportunity to directly know their vectors of decision that can be used within the partnership and, as a result, restrict their own area of freedom (Figure 5.13).

In the case of $CAD = \emptyset$, partners are not interoperable, i.e. they are unable to make a decision that satisfies both the objectives of the partnership and their own interests. However, decision makers can modify their decisional capacity, if possible, in order to define feasible solutions.

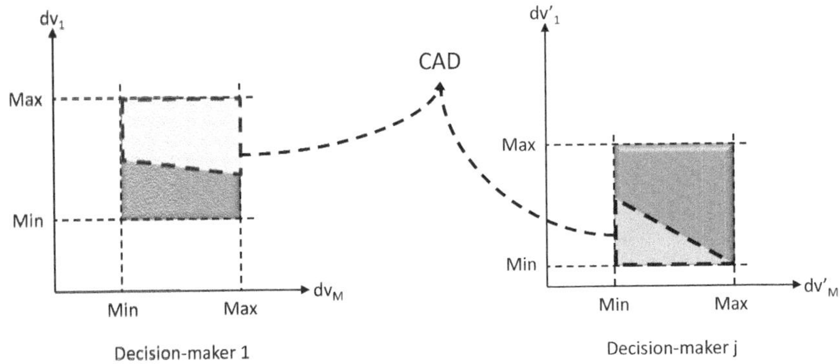

Figure 5.13. *Common area of decision and mapping with the decision frames of partners*

To illustrate the decisional interoperability approach, let us consider the following simple example. Two partners, respectively E1 and E2, work in parallel on the production of products which will be assembled, later, to form a final device (assembling is not considered in this example). The objectives of the partnership regarding the production of these products are: (1) "*to respect a maximum cost of 30K€*" and (2) "*to respect a production time of 15 days*". Figure 5.13 shows the link between the decisional interoperability approach and formalization and the proposed example.

First, the partners have to establish their decisional capacity, i.e. their admissible limits regarding the decision frame used in the context of the partnership. From their decision frames, partners extract a set of possible decisions and their associated performances with regards to the objectives of the partnership (*2-admissible decisions*).

Then, feasible solutions within the partnership, i.e. the vectors of values supported only by the decisional capacities of each partner, are defined. The solutions that will allow (1) reaching the objectives of the partnership and (2) respecting the decisional capacity of each partner constitute a set from which achievable performance can be obtained (*3-Feasible solution*). These performances are made with dedicated functions (*4-Function*). For instance, the performance criterion of the second objective of the partnership is the time. As partners perform activities in parallel, the function is a "MAX" of vectors of values (related to the objective) for each feasible solution.

Finally, the common area of decision is defined as the set of feasible solutions that allows reaching both the objectives of the partnership and the interests of the partners (*5-Common area of decision*). Thereafter, decision makers can restrain their own decision frames by excluding decisions that do not allow achieving objectives. In the following example, the decision frames of partners will not be modified since each extracted feasible solutions.

	Formalisation	Application
1-Objectives	(ob'_1, ob'_2)	ob'_1: Respect a maximum cost (30 K€) ob'_2: Respect a lead time (15 days)
2-Admissible decisions	For partner 1: (d_{11}, d_{12}, d_{13}) For partner 2: (d_{21}, d_{22}, d_{23})	$(10, 15, 17)$ for ob'_1 $(14, 7, 2)$ for ob'_2 $(13, 16, 20)$ for ob'1 $(15, 12, 9)$ for ob'2
3-Feasible solutions	$R = \left\{ r_n \middle\vert r_n = \begin{bmatrix} d_{1k} \\ d_{2k} \end{bmatrix} \right\}$	$r_1 = \begin{bmatrix} DV_{11} \\ DV_{21} \end{bmatrix}, r_2 = \begin{bmatrix} DV_{11} \\ DV_{22} \end{bmatrix}, r_3 = \begin{bmatrix} DV_{11} \\ DV_{23} \end{bmatrix}$ $r_4 = \begin{bmatrix} DV_{12} \\ DV_{21} \end{bmatrix}, r_5 = \begin{bmatrix} DV_{12} \\ DV_{22} \end{bmatrix}, r_6 = \begin{bmatrix} DV_{12} \\ DV_{23} \end{bmatrix}$ $r_7 = \begin{bmatrix} DV_{13} \\ DV_{21} \end{bmatrix}, r_8 = \begin{bmatrix} DV_{13} \\ DV_{22} \end{bmatrix}, r_9 = \begin{bmatrix} DV_{13} \\ DV_{23} \end{bmatrix}$
4-Function	$ob'_1: f_1(r_n) = \sum_{j=1,k=1}^{P,M} d_{jk}$ $ob'_2: f_2(r_n) = MAX_{j=1,k=1}^{P,M} d_{,k} \leq T_{max}$	$f(r_1) = (23, 15), f(r_2) = (26, 14), f(r_3) = (23, 15)$ $f(r_4) = (28, 15), f(r_5) = (31, 12), f(r_6) = (35, 9)$ $f(r_7) = (30, 15), f(r_8) = (33, 12), f(r_9) = (37, 9)$
5-Common area of decision	CAD = $\{r_1, r_2, r_3, r_4, r_7\}$	CAD ={(23, 15), (26, 14), (23, 15), (28, 15), (30, 15)}

Figure 5.14. *Formalization of the concept of decisional interoperability and application*

This chapter demonstrated the principles, the concept and their formalization to realize decisional interoperability. It is mainly based on the clear understanding and the clear definition of information that have to be shared by several decision makers in order to make a decision. This has to lead to a better organization of the decision-making activity and a reduction of time to make a decision, and decision makers can also rely on existing tools (related to mechanisms between different partners; for instance, game theory) to build most the efficient decision.

However, this kind of approach – strongly based on the involvement of actors – should not hide the grim industrial reality. On the one hand, a network of enterprises includes more than two partners (in comparison with the illustration previously presented). Thus, it can be time consuming and difficult to find a solution that satisfies both the partnership and each partner. Decisional interoperability can help overcome these problems (numerous iterations, impact of a decision taken by a decision maker on other decision makers, etc.). On the other hand, information shared within the network can be incorrect (deliberately or not), not visible (restriction, partial) and this can be harmful for the partnership. Decisional interoperability does not consider this aspect and decision makers will have to consider, additionally, decision-making approaches in the context of uncertainty. As a result, even though decisional interoperability provides the basic principles to work together and share information, decision makers have to use available tools for decision-making.

5.4. Conclusion

This chapter addressed the problem of the development of decisional interoperability. Decisional interoperability is established as an organizational approach by using and providing adequate information to share in order to make decisions. This approach is independent from the technical aspect, can be easily implemented and can facilitate and improve decision-making, rapidly. Furthermore, the act of highlighting a common area of decision prior to any decision and building on the basis of a set of information to exchange (provided by partners) can allow solving and preventing classical problems identified in the collective decision-making activity.

The proposed approach is based on the concepts developed in the GRAI decisional model for enterprise integration from the decisional point of view. The full understanding of these concepts has enabled us to extend and to adapt those to the context of interoperability.

5.5. Bibliography

[BEN 12] BENS I., *Facilitating with Ease!: Core Skills for Facilitators, Team Leaders and Members, Managers, Consultants, and Trainer*, 3rd ed., John Wiley & Sons, New York, 2012.

[CEN 04] CEN, Enterprise Integration – Decisional Reference Model, CEN TS 14818, TC310/WG1, 2004.

[CHE 05] CHEN D., Practices, principles and patterns for interoperability, INTEROP Deliverable D6.1, 2005.

[DAC 06] DACLIN N., CHEN D., VALLESPIR B., "Decisional interoperability: concepts and formalisation", *Proceedings of 7th IFIP Working Conference on Virtual Enterprises (PROVE'06)*, Helsinki, Finland, vol. 224, pp. 297–304, 25–27 September 2006.

[DAC 07] DACLIN N., Contribution au développement d'une méthodologie pour l'interopérabilité des entreprises, PhD Thesis, University of Bordeaux 1, December 2007.

[DHA 05] DHAENENS-FLIPO C., "Optimisation combinatoire multi-objectif: apport des méthodes coopératives et contribution à l'extraction de connaissances", Mémoire d'Habilitation à Diriger les Recherches, Lille University of Science and Technology, 5 October 2005.

[DOU 98] DOUMEINGTS G., VALLESPIR B., CHEN D., "GRAI Grid Decisional Modelling", in BERNUS P., MERTINS K., SCHMIDT G. (eds), *International Handbook on Information Systems*, Springer, Berlin, 1998.

[END 95] ENDSLEY M.R., "Toward a theory of situation awareness in dynamic systems", *Human Factors*, vol. 37, no. 1, pp. 32–64, 1995.

[HAM 03] HAMMAMI A., Modélisation technico-économique d'une chaîne logistique dans une entreprise réseau, PhD Thesis, Ecole Nationale Supérieure des Mines de Saint-Etienne, 2003.

[IMO 14] IMOUSSATEN A., MONTMAIN J., MAURIS G., "A multicriteria decision support system using a possibility representation for managing inconsistent assessments of experts involved in emergency situations", *International Journal of Intelligent Systems*, vol. 29, pp. 50–83, 2014.

[ISO 11] ISO, Advanced Automation Technologies and Their Applications – Part 1: Framework for Enterprise Interoperability, ISO/TC 184/SC 5, 2011.

[KVA 00] KVAN T., "Collaborative design: what is it?", *Automation in Construction*, vol. 9, pp. 409–415, 2000.

[MON 03] MONSARRAT E., BRIAND C., ESQUIROL P., "Une aide à la cooperation inter-entreprise pour la production à la commande", *Proceedings of Coopération Innovation et Technologies (CITE'2003)*, Troyes, France, pp. 23–36, 3–4 December 2003.

[MON 15] MONTMAIN J., LABREUCHE C., IMOUSSATEN A. *et al.*, "Multi-criteria improvement of complex systems", *Information Science*, vol. 291, pp. 61–84, 2015.

[PÈN 06] PÈNARD, T., "Introduction à la théorie des jeux", support de cours de master recherche économie, Faculté de sciences économiques, University of Rennes 1, 2006.

[QUI 04] QUINQUETON J., "Organisation des systèmes multi-agents", support de cours de DEA informatique, University of Montpellier 2, 2004.

[RAS 81] RASMUSSEN J., PEDERSEN O.M., CARNINO A. *et al.*, Classification system for reporting events involving human malfunctions, Report EUR 7444-EN, Commission of the European Communities – Nuclear Science and Technology, 1981.

[SIM 60] SIMON H.A., *The New Science of Management Decision*, Harper and Row Publisher, New York, 1960.

[VAL 06] VALLESPIR B., DOUMEINGTS G., The GRAI method Part 1: global modeling, Lecture Note, INTEROP Network of Excellence, 2006.

[VAL 03] VALLESPIR B., Modélisation d'entreprise et architecture de conduite des systèmes de production, Thesis, University Bordeaux 1, 2003.

[WEI 08] WEISS W.A.R., *An Introduction to Set Theory*, Department of Mathematics, University of Toronto, 2008.

6

The Interoperability Measurement

6.1. Introduction

The changing industrial and economic context leads enterprises to develop impermanent partnership. Therefore, the development of interoperability [ISO 11] between partners has become a major issue over the last two decades, which will continue to remain so in the future. In this context, methods are developed to implement interoperability solutions and evaluate interoperability. From the viewpoint of evaluation, the measure of interoperability allows enterprises to know their strengths and weaknesses in terms of interoperability, and has led to their improvement and capability to avoid deficiencies. Thus, developing interoperability measurement is becoming an important challenge that can help partners to manage and make their collaboration better. Some works have been already done in this domain; however, it remains difficult to define metrics, mainly because of the difficulty of identifying the parameters to characterize interoperability.

In general, interoperability measurement aims at defining metrics to qualify interoperability. The implementation of metrics for interoperability is related to two principles: (1) the identification of the parameters relating to interoperability and (2) the characterization of these parameters by metrics. First, enterprises are not interoperable because barriers to interoperability exist. These interoperability barriers (use of different syntaxes and semantics to represent information, incompatible IT services and infrastructures as well as different organizational structures and methods of works) create obstacles to improve mutual understanding, seemly communicate and transmit data and information and establish collaborative processes. Common barriers exist in all enterprises regardless of the sector of

---

Chapter written by Nicolas DACLIN, David CHEN and Bruno VALLESPIR.

activities and size. These barriers are commonly identified [CHE 11] as *conceptual* (syntactic and semantic), *organizational* (responsibility/authority, organizational structure) and *technological* (related to the computer technology). As a result, interoperability can be seen, at first glance, as a problem of compatibility between two systems, not only at ICT level but also in all levels of the enterprise (business, process, services, data). Thus, developing interoperability means developing knowledge that allows the identification of incompatibilities accurately in order to implement further solutions that remove incompatibilities. Second, defects in a partnership result from poor performances of interaction (interoperation) between partners. This lack of interoperation is related to the exchange of information, services, product, etc., and their exploitation once exchanged. Thus, measuring the performance of interoperation means characterizing the ability of interoperation – in terms of exchange and exploitation – according to the essential criteria for enterprises (e.g. cost, time and quality). To this purpose, a set of clearly defined metrics, enabling partners to know their agility in terms of interoperation, has to be developed. From these considerations, three different types of measurement are considered: (1) the interoperability *potentiality measurement*, allowing a given partner to know their ability to establish interoperability with other partners, (2) the interoperability *compatibility measurement*, allowing for the detection of incompatibilities existing between partners and (3) the interoperability *performance measurement*, allowing for the evaluation of performance of interoperation with regard to the defined criteria.

Applying interoperability measurement in a company helps it to detect possible interoperability problems with respect to its partners (suppliers, customers, subcontractors, etc.). On the basis of the identified interoperability barriers/ problems, the interoperability measurement allows us to characterize the importance of the problems and then propose appropriate solutions to solve the problems and improve interoperability performances (such as interoperability cost, time and quality).

This chapter is organized as follows. Research works within the field of interoperability measurement are presented and discussed in Section 6.2. Section 6.3 outlines the set of interoperability measurement components (potentiality, compatibility and performance measurement) and provides some running examples to better understand the metrics. Section 6.4 discusses the three measurements developed and their possible improvement. The final section presents the conclusion and prospects for interoperability measurement. Note that the interoperability measurement - presented in the following sections – is part of the methodology form enterprise interoperability shown in [DAC 16].

6.2. Models for evaluation of interoperability

Improving interoperability means that metrics to measure it are available. The measure gives information about strengths and weaknesses to interoperate and allows us to prioritize actions to improve ability to interoperate.

Interoperability can be measured from a maturity point of view [C4I 98, KAS 04, GUÉ 13, GUO 11, VAN 12, CAM 13]. The term "maturity" has different meanings in different domains. The most literal definition is "how much you act like an adult". In the enterprise interoperability domain, maturity refers to the ability to adapt and make necessary changes to exchange information/service and use the exchanged information/service. A maturity model defines levels according to a predefined set of interest areas. Maturity is measured by the achievement of the specific and generic goals describing a given level. A maturity model provides recommendations and practices to evolve throughout maturity levels for a progressive improvement and achievement of full maturity (adapted from CMMI). The SEI (Software Engineering Institute) with the Capability Maturity Model (CMM) defines five maturity levels [CMM 04], namely initial, repeatable, defined, managed and optimizing. Other models in different disciplines focus on different levels of the enterprise, for example, the Service-Oriented Architecture Maturity Model [BAC 05], the Extended Enterprise Architecture Maturity Model [IFE 04] and the NASCIO [NAS 03] Enterprise Architecture Maturity Model. These models evaluate processes within organizations (business processes) and identify the best practices which are useful to increase the maturity of their processes and then their level of performance.

The models presented below define maturity levels in terms of interoperability. Some focus on maturity between several systems and others on a single system. Furthermore, some models give recommendations to evolve and to reach full interoperability. The LISI (Levels of Information Systems Interoperability) is one of the first initiatives and defines five levels: isolated, connected, functional, domain and enterprise [C4I 98] to assess the interoperability of information systems. LISI also considers other factors that affect the ability of information systems to interoperate. These factors are categorized into four attributes: procedures, applications, infrastructures and data.

Several similar approaches consider other aspects of interoperability. The OIM (Organizational Interoperability Model) [FEW 03] is an extension of the LISI focused on the organizational viewpoint and according to five levels. Its goal is to extend the LISI models and their levels are aligned for an efficient integration of these models, as presented in [MOR 04].

The LCIM (Levels of Conceptual Interoperability Model) [TOL 07] deals with conceptual interoperability and defines seven levels. The LCIM postulates that interoperability is also a conceptual problem. Thus, this model focuses on the quality of data to exchange. To reach a high level of maturity LCIM recommends the adoption and use of standards.

From a manufacturing enterprise perspective, ATHENA project has developed the EIMM (Enterprise Interoperability Maturity Model) to address interoperability issues at all levels of the company [ATH 05]. Defining the EIMM involves two tasks: (1) identifying the main areas of concern, required to achieve interoperability both internally and externally and (2) defining the maturity levels that describe the improvement path for each area of concern. The EIMM is applied before collaboration, and attempts to cover all fields of industrial systems that are related to interoperability. However, it does not consider conceptual, technological and organizational interoperability.

From an operational performance measurement point of view, [KAS 04], [FOR 08] and [HAM 02], develop metrics for interoperability measurement.

[KAS 04] proposes some metrics (quality attributes) to evaluate interoperability in terms of communication of information. In this sense, several equations (connectivity, capacity, overload, underutilization, undercapacity) to evaluate operational interoperability are provided (e.g. overload metric to measure if the amount of data to be exchanged is higher than a system is able to transmit). Ford proposes seven modes related to operational interoperability effectiveness and their associated metrics: directional, auto, pure, contextual, time-variant, constrained upper bound and collaborative/confrontational. For instance, directional mode considers either bidirectional (partner 1 ←→ partner 2) or unidirectional (partner 1 → partner 2) interoperability. [HAM 02] proposes a method based on the operational requirements definition (e.g. operational readiness and speed exchange) and evaluates whether systems meet interoperability requirements. A table allows us to evaluate the level (color code) of satisfaction of interoperability requirements. Finally, other works are related to interoperability measurement such as the Service Level Agreement (SLA) [VER 99] that measures the Quality of Services (QoS) between a provider and a customer or else, [YAH 12], proposes some formal measures for semantic interoperability.

Some maturity models are based on the classical categorization of interoperability, that is, conceptual, organizational and technological. However, no general model to gather and evaluate all the categories is developed. Furthermore, these models consider only the maturity and not the operational interoperability. Therefore, it is difficult to measure and evaluate the operational performances related to the exchange and the compatibility. In this way, it is also required to propose tools to measure interoperability during interoperations. Thus, the following four main characteristics have to be considered to develop methods to measure interoperability: (1) categories of interoperability, (2) intra-enterprise maturity evaluation, (3) inter-enterprise maturity evaluation and (4) operational performances in terms of interoperability.

6.3. Interoperability measurement

Interoperability measurement can be addressed at two levels [DAC 16]. It can be *ex ante*, i.e. before any collaboration. The evaluation is performed in an intra-enterprise context. This measure is called *potentiality measurement.* It can be considered *in itinere*, i.e. while the collaboration is performed so, the evaluation of interoperability is done in an inter-enterprise context. This measure is named *interoperability degree measurement.*

6.3.1. *The potentiality measurement*

Potentiality means that an enterprise owns intrinsic attributes related to the categories of interoperability (use of standard applications, enterprise models, database, etc.), allowing easy interoperability in the eventuality of a partnership [DAC 08]. Potentiality is an intra-enterprise evaluation without knowing the interoperating partner. It aims to increase the ability to implement interoperability and decrease the possibility of facing up to problems during a future partnership. The potentiality model is defined according to three interoperability problems and the levels where interoperability takes place. For each problem and each level, five potentiality levels are defined: (1) *isolated*, means a total incapacity to interoperate; (2) *initial*, means interoperability requires strong efforts that affect the partnership; (3) *executable*, means interoperability is possible even if the risk of encountering problems is high; (4) *connectable*, means interoperability is easy even if problems can appear for distant partnership and (5) *interoperable*, which considers the evolution of levels of interoperability, and where the risk of meeting problems is weak. Figure 6.1 presents the attributes – for each category of interoperability – that a given enterprise has to respect to reach the interoperable level.

	Conceptual	Technological	Organizational
Interoperable	• Semantic and syntax (of models) are fully and well defined • Evolution of semantic and syntax are considered	• Standards are used as technical environment • Environment is maintained and updated according to technical evolution	• Models, responsibility, authority and skills are known explicitly and updated after evolution • Outsourceable information is grouped and updated in dedicated database • Access privileges to this database are defined and updated
Connectable	• Semantic and syntax (of models) are fully and well defined.	• Standards are used as technical environment	• Models, responsibility, authority and skills are known explicitly • Outsourceable information is grouped in dedicated database • Access privileges to this database are defined
Executable	• Semantic and syntax (of models) are known but not fully defined.	• Technical environment is existing • Communication protocol is used to exchange information	• Models, responsibility, authority and skills are known explicitly • Outsourceable information is only known
Initial	• Semantic and syntax (of models) are known but not explicitly defined	• Technical environment is existing •No communication protocol is used to exchange information	• Models, responsibility, authority and skills are only known • Outsourceable information is unknown
Isolated	• Semantic and syntax (of models) are unknown	• No technical environment is existing • Exchange of information can be performed only in a manual way	• Models are not defined • Responsiblility authority and skills are not identified

Figure 6.1. *An enterprise's potentiality*

In the end, the goal is to reach the top level. Although this evolution obviously decreases the risk of problems and facilitates the implementation of interoperability, maximum potentiality is not full interoperability. Indeed, the use of standard tools by an enterprise does not ensure that a partner will use the same ones and problems of interoperability can still occur. For instance, the use of BPMN 2.0 [OMG 11] – to model business processes – does not mean that a future partner uses this language. Nevertheless, it is very likely that other enterprises use it and it can be linked to other languages easily (richness of modeling objects, use of xmi to describe model). Furthermore, and even if partners are using the same modeling language, it is possible that partners are not yet interoperable. On the one hand, other aspects of interoperability have to be considered (e.g. technological) on the other hand, interoperability has to be assessed during the running phase of the partnership.

6.3.2. *Interoperability degree measurement*

Interoperability degree gives information to partners about the presence (or absence) of incompatibilities disturbing the interoperability. This evaluation allows partners to avoid and/or correct deficiencies during the collaboration. The interoperability degree includes compatibility measurement and performance measurement.

Compatibility measurement is donc during the engineering stage, that is, when systems need to set up interoperability with a known partner. This measurement is performed when the partner/system of the interoperation is known. The measurement considers the identified problem of interoperability.

According to interoperability views and interoperability problems, the objective is to highlight incompatibilities of which the following questions can be asked to detect incompatibility between two systems. The set of questions is selected according to the interoperability expectations expressed by partners in terms of enterprise view where interoperability has to be developed.

– *Conceptual compatibility for data view*:

- *Syntactic:* Is the data to be exchanged expressed with the same data model? Does the data to be exchanged use a data model?

- *Semantic:* Does the data to be exchanged have the same meaning? Does the meaning of exchanged data need to be explicitly defined?

– *Organizational compatibility for process view*:

- *Process authorities:* are processes of authorities/responsibilities clearly defined on both sides?

- *Organization structures:* are the organizational structures of processes expressed on both sides?

– *Technological compatibility for business view*:

- *Platform:* are the IT platform technologies compatible? Do partners use the IT platform?

- *Communications:* do partners use the same protocols of exchange? Do partners use protocol of exchange?

Following this rule, the compatibility matrix shown in Figure 6.2, can be built. If an incompatibility is detected, coefficient 1 is assigned to the interoperating level (view) and problem. Conversely, coefficient 0 is applied either when no incompatibility is detected or when the view is not concerned by the development of interoperability or others, when it is considered useless (for instance, partners do not want to make their processes interoperable put coefficient 0 at each intersection between the process level and interoperability problems). Binary values allow us to identify easily and rapidly which enterprise level, and category of interoperability, are concerned by incompatibilities. These values are an indication for the existence (or not) of a problem, and do not reflect its importance. For instance, an application incompatibility does not state if applications are fully incompatible or if some modules are compatible. Partners have to identify clearly and precisely the problem to solve.

	Conceptual		Organizational		Technological	
	Syntactic	Semantic	Authorities responsibilities	Organization	Platform application	Communication
Business	0	0	0	0	**1**	0
Process	0	0	**1**	**1**	0	0
Services	0	0	0	0	0	0
Data	0	**1**	0	0	0	0

Figure 6.2. *The compatibility measurement matrix*

The highest degree of compatibility means that all incompatibilities are removed. The inverse situation means the lowest degree of compatibility. The compatibility measure allows us to know the incompatibilities. The higher the number of incompatibilities, the more important the efforts of the partners to become interoperable.

In a similar way, the incompatibility measurement allows us to prioritize the actions to improve interoperability and define a migration path to follow and remove the identified incompatibilities in a progressive way. Partners have to perform their migration path and actions jointly, to ensure that all barriers are really removed for each of them.

Finally, the overall compatibility degree is obtained by the summation of all the coefficients of compatibility, as shown in equation [6.1]:

$$d_{comp} = \sum_{i=1,j=1}^{n,6} coef_{comp\, i,j}, i, j, n \in \mathrm{N}^* \qquad [6.1]$$

where:

– d_{comp} is the compatibility degree. A compatibility degree equal to 0 means no problem of compatibility. A compatibility degree equal to 24 means interoperability is developed at each level, for each category, but incompatibilities are present;

– $coef_{comp}$ is the coefficient of compatibility. It is related to each view and sublevels of syntactic, semantic, authorities, responsibility, application, communication categories). A coefficient is equal to 0 or 1, 1 meaning incompatibility;

– i represents the views where interoperability is implemented;

– j represents the sublevels of categories.

As an example, let us consider an exchange of data related to a product. As Figure 6.3 (upper part) shows, partners use (1) different words to name and describe their product (e.g. product vs. item and name vs. designation) and (2) two different data models (e.g. xml vs. EXPRESS). This can lead to a misunderstanding (semantic problem) and an impossibility to use the information exchanged (syntactic problem). As a result, two incompatibilities are identified (lower part) at the data level for the conceptual category.

Partner 1

```
<?xml version="1.0" encoding="ISO-8859-1"?>
<product>
        <name> STRING </name>
        <id> INTEGER </id>
</product>
```

Partner 2

```
SCHEMA item_schema;

ENTITY Item
        designation: STRING;
        ref: INTEGER;
END_ENTITY;

END_SCHEMA;
```

	Conceptual		Organizational		Technological	
	Syntactic	Semantic	Authorities responsibilities	Organization	Platform application	Communication
Business	0	0	0	0	0	0
Process	0	0	0	0	0	0
Services	0	0	0	0	0	0
Data	**1**	**1**	0	0	0	0

Figure 6.3. *Example of incompatibilities at the data level (semantic and syntactic)*

After that, the interoperability framework is used to find the most relevant solutions. Figure 6.4 presents some identified interoperability solutions positioned framework. According to the interoperability views and category of interoperability. For instance, semantic annotation or ontology solve conceptual interoperability for each interoperability view and are fully adapted to implement interoperability from a unified approach. In the previous example there were two incompatibilities at the data level for the conceptual category (semantic and syntactic). If the partners choose to remove incompatibilities based on a unified approach (left-hand side on Figure 6.2), the framework makes a set of solutions that meets their expectations available. Thus, for data interoperability level and conceptual category, the semantic annotation is solution. A complete set of interoperability solutions in agreement with the enterprise interoperability framework can be found in INTEROP [INT 06a].

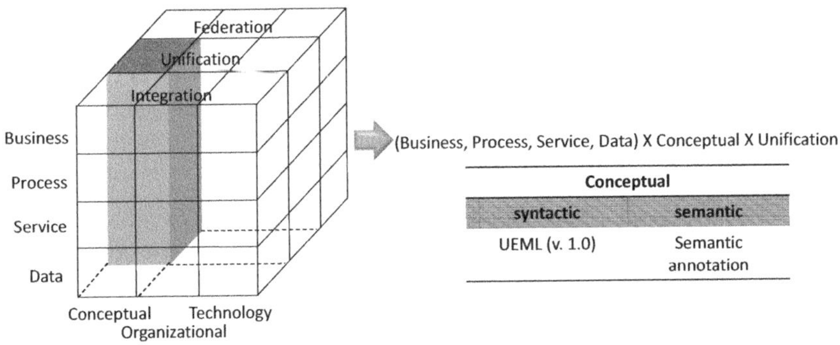

Figure 6.4. *Enterprise interoperability framework and examples of interoperability solutions*

6.3.3. *Performance measurement*

The performance measurement of interoperability allows enterprises to evaluate interoperability according to a set of predefined criteria, such as:

– criteria are defined to measure and evaluate the impact of the implementation of interoperability on enterprise (e.g. in terms of costs). In this case, enterprises evaluate the potential necessity to apply interoperability;

– criteria are defined and related to the characteristics of interoperability (e.g. communication, exploitation), and only their absolute value are measured. If the measures are not related to objectives defined by partners, it is difficult to evaluate the importance of deficiencies and to initiate corrective actions;

– criteria are defined and related to the characteristics of interoperability, the absolute value is measured and the result is compared to objectives. In this case, it is possible to evaluate precisely the achievement of the objectives and to initiate corrective actions if a drift accurs.

The measure of the *performance of interoperation* focuses on the third type and is performed during the operational phase, that is, runtime, to evaluate the ability of interoperation between two enterprises and is based on the interoperation cycle (Figure 6.5) including (1) a phase of operation (p_1 side), (2) a phase of exchange and (3) a phase of operation (p_2 side). The first phase is related to the operation on data, models, decision frame, etc. to exchange or share. "Operation$_1$" means that a given partner – known as "sender" – has the set of data, models, etc. requested by an interoperation partner, ready to be sent. The second phase is related to the exchange or sharing of data, models, decision frame, etc. "Exchange" means that the "sender" – sends (regardless of the support), or makes available, data, models etc. for the interoperation partner. The third phase relates to the operation of the exchanged or shared data, models, decision frame, etc. by the partner receiver. "Operation$_2$" means that the "receiver" – is able to operate the data, models, etc. provided by a sender. For instance, a receiver is able to understand and process received data to perform a given task.

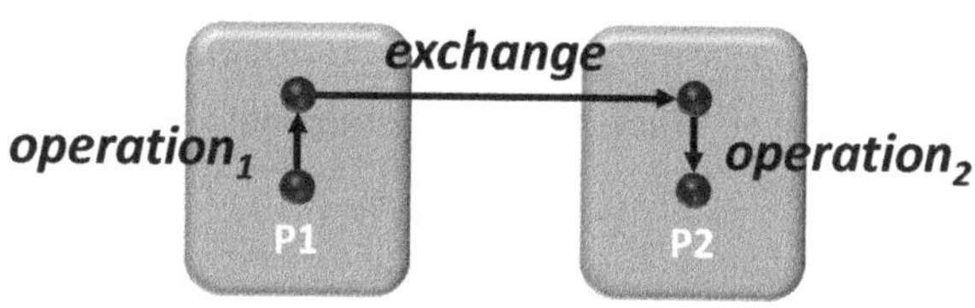

Figure 6.5. *The interoperation cycle (operation$_1$ – exchange – operation$_2$)*

Classical criteria such as cost, time and quality are used to measure performance. The following parts present each criterion and an example to illustrate their application. Let us consider interoperation, a partner P1 (sender) that sends a set of four data (e.g. parts list, routing sheet, code, plan) requested by a partner P2 to perform their own tasks (e.g. Computer Numeric Control machining). These data are sent via e-mail, and the partnership is limited to one interoperation cycle (i.e. 1 operation$_1$, 1 exchange and 1 operation$_2$).

The cost of interoperation (C_{in}) represents the cost engaged to perform an interoperation cycle. It is made of the cost of exchange (C_{ex}) and the cost of operation (C_{op}). The evaluation of the cost of interoperation, corresponds to the comparison of the real cost of exchange and operation with the cost of exchange and operation that enterprises wish to engage. If one of these real costs is greater than the expected costs, then there is a deficiency. The cost of exchange and the cost of operation are then expressed such as:

$$C_{in} = C_{ex} + C_{exp} \quad [6.2]$$

$$C_{ex} = C_{iniex} - C_{effex} \quad [6.3]$$

$$C_{op1} = C_{iniop1} - C_{effop1} \quad [6.4]$$

$$C_{op2} = C_{iniop1} - C_{effop2} \quad [6.5]$$

where:

– C_{in} is the overall cost of interoperation. If $C_{in} < 0$, the criterion is not respected (partners have to correct incriminated criteria). Conversely, $C_{in} \geq 0$ indicates that this criterion is respected (a given criterion can offset the other, and partners can accept this statement);

– C_{ex} is the cost of exchange. It represents the difference between the initial cost allocated to exchange and the real cost of exchange;

– C_{iniex} is the initial cost allocated for exchange;

– C_{effex} is the real cost of exchange;

– C_{op} is the cost of operation. It represents the difference between the initial cost allocated to operation and the real cost of operation;

– C_{iniop} is the initial cost allocated for operation;

– C_{effop} is the real cost of operation.

Figure 6.6 shows the calculation of the cost of exchange and the cost of operation.

Performance	Details	Description
Cost of operation$_1$	$C_{op1} = 0$	C_{op1} = 0, cost of preparation of information to send is considered as negligible by the sender
Cost of exchange	$C_{ex} = 0$	C_{ex} = 0, cost of emailing is considered as negligible by the sender
Cost of operation$_2$	$C_{op2} = C_{iniop2} - C_{effop2} = 0$	C_{iniop} = €380, the receiver has allocated this cost to exploit data C_{effop} = €380, order has to be transformed to be exploited by software used by receiver. An IT Services & Software Engineering company has performed the transformation sheets (charged €400/day)

Figure 6.6. *Example of the calculus of cost of interoperation*

This example considers that the cost of operation for partner 1 and the cost of exchange will be equal to 0 for each interoperation, and the cost of operation for partner 2 is engaged only for the first interoperation (transformation sheet is then used for each interoperation).

The duration of interoperation (D_{in}) corresponds to the duration between the date at which information starts to be operated for sending and the date at which the requested information is effectively exploitable. The principle is similar to the one applied for the cost. The duration of interoperation is broken down into two intervals: the duration of exchange (D_{ex}) and the duration of operation (D_{op}). The evaluation of the duration of interoperation corresponds to the comparison of the real duration of exchange and operation with the duration of exchange and operation expected by the partners. If none of these durations is greater than those expected, there is no deficiency and the durations of exchange and operation are correct. The equations given below represent the duration of interoperation according to this decomposition (adapted from [KAS 04]):

$$D_{in} = D_{ex} + D_{op1} + D_{op2} \quad [6.6]$$

$$D_{ex} = D_{iniex} - D_{effex} \quad [6.7]$$

$$D_{op1/2} = D_{iniop1/2} - D_{effop1/2} \quad [6.8]$$

$$D_{effex} = T_{rec2} - T_{em1} \quad [6.9]$$

$$D_{effop1} = T_{em1} - T_{sop1} \quad [6.10]$$

$$D_{effop2} = T_{op2} - T_{rec2} \quad [6.11]$$

where:

– D_{in} is the overall duration of interoperation. If the duration of $D_{in}<0$, the criterion is not respected (partners have to correct incriminated criteria). Conversely, if the duration of interoperation is $D_{in}\geq 0$, the criterion is respected (a given criterion can offset the other and partners can accept this statement);

– D_{ex} is the duration of exchange. It represents the difference between the initial duration allocated to exchange and the real duration of exchange. If $D_{ex}<0$, the criterion is not respected (effective duration of exchange is higher than the initial duration of exchange). Conversely, if $D_{ex}\geq 0$, the criterion is respected;

– D_{iniex} is the initial duration expected for exchange;

– $D_{op1/2}$ is the duration of operation (partner 1/2). It represents the difference between the initial duration expected to operation and the real duration of operation. If $D_{op}<0$, the criterion is not respected (effective duration of operation is higher than the initial duration of operation). Conversely, if $D_{op}\geq 0$, the criterion is respected;

– $D_{iniop1/2}$ is the initial duration expected for operation 1/2;

– D_{effex} is the real duration of exchange. It represents the time measurement between the date of emission of information (partner 1) and the date of reception of the information (partner 2);

– D_{effop1} is the real duration of operation 1. It represents the time measurement between the date at which partner 1 starts to prepare the set of information to send and the date at which this set is ready to be sent;

– D_{effop2} is the real duration of operation 2. It represents the time measurement between the date of the reception of information and the date of operation, that is, information can be really exploited;

– T_{sop1} is the date at which partner 1 starts to operate information to send;

– T_{rec2} is the date of reception by partner 2;

– T_{em1} is the date of emission by partner 1;

– T_{op2} is the date of operation by partner 2.

Figure 6.7 presents the evaluation of the duration of interoperation and its terms.

Performance	Details	Description
Duration of operation$_1$	$D_{op1} = D_{iniop1} - (T_{em1} - T_{sop1}) = 0min$	D_{iniop} = 10 min, the expected duration of operation$_1$ T_{em1} = 11.40am, the sender starts to transmit at this date T_{sop1} = 11.30am, the sender starts to operate at this date
Duration of exchange	$D_{ex} = D_{iniex} - (T_{rec2} - T_{em1}) = 0$ min	D_{iniex} = 10 min, the expected duration of exchange T_{rec2} = 11.50am, the receiver receipts data at this date T_{em1} = 11.40am, the sender starts to transmit at this date
Duration of operation$_2$	$D_{op2} = D_{iniop2} - (T_{op2} - T_{rec2}) = 0$ min	D_{iniop2} = 10 min, the expected duration of operation$_2$ T_{op2} = 12.00am, after treatment the receiver can exploit data at this date T_{rec2} = 11.50am, the receiver receipts data at this date

Figure 6.7. *Example of the calculus of duration of interoperation*

The quality of interoperation (Q_{in}) considers four types of quality: (1) quality of operation 1, (2) quality of exchange, (3) quality of operation 2 and (4) conformity. The evaluation of the quality of interoperation corresponds to the comparison of the real quality of exchange, operation and conformity with the quality of exchange, operation and conformity expected by the partners. These types of quality are expressed by the following equations:

$$Q_{in} = Q_{ex} + |Q_{op1}| + |Q_{op2}| + Q_{conf} \quad [6.12]$$

$$Q_{ex} = Q_{iniex} - Q_{effex} \quad [6.13]$$

$$Q_{op1/2} = Q_{iniop1/2} - Q_{effop1/2} \quad [6.14]$$

$$Q_{conf} = Q_{iniconf} - Q_{effconf} \quad [6.15]$$

$$Q_{effop1} = {N_{rdy}}/{N_{req}} \quad [6.16]$$

$$Q_{effex} = {N_{succ}}/{N_{eff}} \quad [6.17]$$

$$Q_{effop2} = {N_{rec}}/{N_{req}} \quad [6.18]$$

$$Q_{effconf} = {N_{conf}}/{N_{rec}} \quad [6.19]$$

where:

– Q_{in} is the overall quality of interoperation. If the quality of interoperation is greater than the expected quality of interoperation, then the criterion is not respected (partners have to correct incriminated criteria). If the quality of interoperation is less than or equal to the expected quality of interoperation, then the criterion is respected;

– Q_{ex} is the quality of exchange. It represents the difference between the initial quality expected to exchange and the real quality of exchange. If the quality of exchange is different from 0, the criterion is not respected (effective quality of exchange is less than the initial quality of exchange). Conversely, if the quality of exchange is equal to 0, the criterion is respected;

– $Q_{op1/2}$ is the quality of operation. It represents the difference between the initial quality expected to operate and the real quality of operation. If the quality of operation is different from 0, the criterion is not respected (effective quality of operation is higher or less than the initial quality of operation). Conversely, if the quality of operation is equal to 0, the criterion is respected;

– Q_{conf} is the quality of conformity. It represents the difference between the initial quality of conformity expected to operate and the real quality of conformity. If the quality of conformity is different from 0, the criterion is not respected (effective quality of conformity is less than the initial quality of conformity). Conversely, if the quality of conformity is equal to 0, the criterion is respected;

– Q_{iniex} is the initial expected quality of exchange;

– $Q_{iniop1/2}$ is the initial expected quality of operation;

– $Q_{iniconf}$ is the initial expected quality of conformity;

– Q_{effex} is the effective quality of exchange. It draws up if the exchange is correctly performed, that is, if information sent to a partner succeeds. It is expressed by the relation between the number of successful sendings and the total number of sendings. From a strict information system interoperability perspective, it is important to note that Q_{effex} can be used to evaluate the connectivity index proposed by [KAS 04], since this criterion evaluates simply the relation between "*messages transmitted by a node*" and "*messages received by a node*".

– N_{eff} is the number of effective sendings;

– N_{succ} is the number of sendings successfully completed;

– $Q_{effeop1}$ is the effective quality of operation 1. It represents the number of information requested and the number of information that the partner is able to send. It is expressed by the relation between the number of information ready to be sent and the number of requested information;

– Q_{effop2} is the effective quality of operation 2. It represents the number of receptions by a partner by comparison with the number of requests. For instance, a higher amount of receptions (difficulty to process all the information) or a lower amount (shortage of information) of the number of requests means a deficiency. It is expressed by the relation between the number of receptions and the number of requests;

– N_{req} is the number of requests;

– N_{rdy} is the number of information ready to be sent;

– N_{rec} is the number of receptions;

– $Q_{effconf}$ is the effective conformity. It corresponds to the operation of the information, that is, if the fact that the information received is directly exploitable or not. It is expressed by the relation between the number of conformed receptions and the total number of receptions, that is, workable without intermediary;

– N_{conf} is the number of conformed receptions.

Finally, the evaluation of the quality of interoperation requires comparing each real quality criterion with the quality criterion expected by partners. If none of these quality levels is different with the expected one, there is no deficiency. Otherwise, the deficient criterion has to be identified to start corrective actions.

Figure 6.8 shows the evaluation of the quality of interoperation and its terms.

Performance	Details	Description
Quality of exchange	$Q_{ex} = Q_{iniex}$-$(N_{succ}/N_{eff}) = 0$	Q_{iniex} = 1/2, the number of failings allowed to exchange N_{eff} = 4, sender has performed 4 sendings N_{succ} = 2, sender has asked for a confirmation of reception
Quality of operation$_1$	$Q_{op1} = Q_{iniop1}$- $(N_{rdy}/N_{req}) = 0$	Q_{iniop1} = 1, the expected quality of operation$_1$ N_{req} = 2, the receiver has requested 2 data N_{rdy} = 2, the sender is able to send 2 data
Quality of operation$_2$	$Q_{op2} = Q_{iniop2}$- $(N_{rec}/N_{req}) = 0$	Q_{iniop2} = 1, the expected quality of operation$_2$ N_{req} = 2, the receiver has requested 2 data N_{rec} = 2, the receiver has received 2 data
Quality of conformity	$Q_{conf} = Q_{iniconf}$-(Nrec- N_{conf}) = 0	$Q_{iniconf}$ = 1, the expected quality of conformity N_{rec} = 2, the receiver has received 2 data N_{conf} = 2, the data are conform to those expected

Figure 6.8. *Example of the calculus of quality of interoperation*

The performance assessment allows us know the global degree of performance in agreement with defined criteria. If a deficiency is detected, coefficient 1 is assigned to the performance attribute. Conversely, coefficient 0 is applied. Following this rule of interpretation, the vector of interoperation performance is built (Figure 6.9). This example shows no gaps in the interoperation performance criteria, except on criterion D_{ex}. For instance, because of a capacity restriction, the sender performed two sendings, leading to a duration of exchange (50 min, preparation of folders and sending) greater than expected. Once again, binary values are easily readable and understandable, but do not give any information about the importance of the gap. Thus, the incriminated criteria has to be clearly located and fully described.

C_{in}			D_{in}			Q_{in}				P_{in}
C_{ex}	C_{op1}	C_{op2}	D_{ex}	D_{op1}	D_{op2}	Q_{ex}	Q_{op1}	Q_{op2}	Q_{conf}	
0	0	0	1	0	0	0	0	0	0	1

Figure 6.9. *The vector of interoperation performance*

The highest degree of performance (measures match with expected results) means that no deficiency appears during the collaboration. The inverse situation (at least one measure differs from the expected results) meaning that deficiency(ies) exist(s) between partners.

The overall degree of performance represents the summation of all the coefficients of performance, as shown in equation [6.13]:

$$p_{int} = \sum_{att=0}^{10} coef_{att\ j} \quad [6.20]$$

where:

– p_{int} represents the performance of interoperation between partners;

– $coef_{att}$ is the coefficient assigned to the performances attributes. A coefficient is equal to 0 or 1, 1 meaning a deficiency on the given attribute;

– *att* represents the attributes of performance that are concerned (C_{ex}, C_{opl}, D_{ex}, etc.).

It is important to note that the performances can be measured in accordance with pre-agreed criteria. For example, SLAs [VER 99, DES 10] that define a kind of contract expressing objectives can be used to evaluate interoperability. These objectives represent the targeted values of performances criteria to evaluate their interoperation (e.g. duration of exchange) and these contracts are negotiated before the interaction. As such, the required quality, time and cost can be agreed between two entities. Using SLAs in a company can indeed improve its interoperability potential and maturity, but this must be carried out before evaluating the performance.

6.4. Taking it further

The metrics in the compatibility matrix are binary (1 expresses an incompatibility is 0 otherwise). The use of binary values is "straightforward"; nevertheless, the main objective is to highlight interoperability problem whether complete or partial. For instance, partially modeled data can be exchanged and exploited and coefficient 1 assigned to the compatibility matrix without knowing either its potential impact or its importance on interoperability. Hence, it is necessary to identify precisely which data are concerned and use a solution to improve the situation (e.g. semantic annotation, ontology [INT 06b]). However, compatibility could be to know the importance of a problem and to further prioritize the implementation of a solution for a given problem in comparison with other identified problems. In this sense, qualitative reasoning [GON 04] and further the fuzzy sets [ZAD 78, BAN 04] can be promising approaches. Nonetheless, the use of fuzzy sets need some further investigations and research. For instance, fuzzy sets require the definition of membership functions and preference models (according to information given by enterprises to extract weights and interactions that are specific to each enterprise).

In any case (1) each application of the metrics is specific to the partnership (and partners) and depends strongly on the expectations and (2) each identified problem must be solved. Thus, it is preferable, at first, to adopt the pair {0, 1} and let partners refine their evaluation when they evaluate their compatibility. Nevertheless, the adoption of a complex multi-level evaluation scale (such as the fuzzy set method) need not reconsider the metrics and their underlying concepts will remain valid.

6.5. Conclusion and prospects

The measurement of interoperability is an important issue for partners that want to interoperate efficiently. The components presented in this part allow partners to evaluate interoperability precisely and accurately throughout two aspects. On the one hand, a partner can evaluate their own ability to interoperate for future partnership. This is done with the evaluation of the potentiality. On the other hand, partners can evaluate their interoperability during a partnership with the consideration of the compatibility measurement – detection of incompatibilities – and the performance of interoperation according to criteria such as cost, time and quality of interoperation. This is done with the evaluation of the degree of interoperability.

Furthermore, these components can be refined and improved by partners that want to get a more acute evaluation and finally an improvement of interoperability. The potentiality allows an enterprise to evaluate its ability to interoperate with other potential partners. However, no solutions and recommendations to evolve through levels are provided. In this way, it is required to develop these two aspects to guide enterprise to reach full potential.

The interoperability degree focuses only on the compatibility and performances. Others attributes to measure interoperability such as the security aspect to exchange information between partners can be envisaged. Furthermore, it is also interesting to integrate these performance measurements to the global performances of a partnership. It means measuring the impact onto business and verifying and justifying (or not) the necessity to implement interoperability.

Lastly, another aspect is to refine the scaling of values in the compatibility and performance matrix. Boolean values (0 or 1) do not allow us to accurately evaluate the compatibility (compatible or not) and the performance (efficient or not). Thus, it is necessary to propose intermediate values (e.g. 0.25, 0.5, 0.75), to know precisely the degree of compatibility and performance. It will allow us, for instance, to give priority to a given solution or performance rather than the other ones and according to the results of evaluation.

6.6. Bibliography

[ATH 05] ATHENA, "Framework for the establishment and management methodology", ATHENA Integrated Project, Deliverable DA1.4, Work package – A1.4 version 1.0 February 2005.

[BAC 05] BACHMAN J., "A new service-oriented architecture maturity model", available at: http://www.omg.org/, 2005.

[BAN 04] BANA E., COSTA C., CHANGAS M., "A career choice problem: an example of how to use MACBETH to build a quantitative value model based on qualitative value judgements", *European Journal of Operational Research*, vol. 153, pp. 323–331, 2004.

[C4I 98] C4ISR, Architecture Working Group (AWG), Levels of Information Systems Interoperability (LISI), OSD(ASD(C3I)), Washington, D.C., March 1998.

[CAM 13] CAMPOSA C., CHALMETA R., GRANGEL R. *et al.*, "Maturity model for interoperability potential measurement", *Journal of Information Systems Management*, vol. 30, no. 3, pp. 218–234, 2013.

[CHE 06] CHEN D., DASSISTI M., ELVESAETER B., INTEROP Network Of Excellence IS 508011, "Enterprise interoperability-framework and knowledge corpus-advanced report", INTEROP Deliverable DI.2, DI (Domain Interoperability), December 2006.

[CHE 11] CHEN D., "Framework for enterprise interoperability and maturity model", in PANETTO H., BOUDJLIDA N. (eds), *Interoperability for Enterprise Software and Applications*, ISTE Ltd, London and John Wiley & Sons, New York, 2011.

[DAC 08] DACLIN N., CHEN D., VALLESPIR B., "Enterprise interoperability measurement: the potentiality", INCOSE INSIGHT, Special feature – The best of France – Forum académique et robAFIS, 2008.

[DAC 16] DACLIN N., CHEN D., VALLESPIR B., "Developing enterprise collaboration: a methodology to implement and improve interoperability", *Enterprise Information Systems*, vol. 10, no. 5, pp. 467–504, 2016.

[DAN 06] D'ANTONIO F., MISSIKOFF M., "Enterprise modelling and ontologies for interoperability", INTEROP Deliverable D8.2, 2006.

[DES 10] DESAI J., *Service Level Agreements – A Legal and Practical Guide*, IT Governance Publishing, Ely, 2010.

[FEW 03] FEWELL S., CLARK T., "Organisational interoperability: evaluation and further development of the OIM model", *8th International Command and Control Research and Technology Symposium*, Washington D.C., 17–19 June 2003.

[FOR 08] FORD C.T., Interoperability measurement, PhD Thesis, Department of the Air Force Air University, Air Force Institute of Technology, 2008.

[GON 04] GONZALEZ-BAIXAULI B., DO PRADO LEITE J.C.S, MYLOPOULOS J., "Visual variability analysis for goal models", *12th IEEE International Conference on Requirements Engineering (RE)*, pp. 198–207, 2004.

[GUÉ 13] GUÉDRIA W., NAUDET Y., CHEN D., "Maturity model for enterprise interoperability", *Enterprise Information Systems*, vol. 9, no. 15, pp. 1–28, 2013.

[GUO 11] GUO F.B., WANG M.Z., LIAO J.J., "A measurement method of the interoperability of integrated systems based on process net", *Applied Mechanics and Materials*, vol. 88–89, pp. 99–103, 2011.

[HAM 02] HAMILTON J., ROSEN J.D., SUMMERS P.A., "An interoperability road map for C4ISR legacy systems", *Acquisition Review Journal*, vol. 28, pp. 17–31, 2002.

[IFE 04] IFEAD, *Extended Enterprise Architecture Maturity Model (E2AMM)*, support guide version 2.0, Institute for Enterprise Architecture Developments 2006, IEAD, 2004.

[ISO 11] ISO, "Advanced automation technologies and their applications – part 1: framework for enterprise interoperability", International Organization for Standardization, ISO 11354, ISO/TC 184/SC 5, 2011.

[KAS 04] KASUNIC M., ANDERSON W., "Measuring systems interoperability: challenges and opportunities", Software engineering measurement and analysis initiative, Technical note CMU/SEI – 2004 – TN – 003, 2004.

[MOR 04] MORRIS E., LEVINE L., MEYERS C. *et al.*, "System of Systems Interoperability (SOSI): Final Report", Technical report CMU/SEI – 2004 – TR – 004, 2004.

[NAS 03] NASCIO, NASCIO Enterprise Architecture Maturity Model, Version 1.3, National Association of State Chief Information Officers, Lexington, December 2003.

[OMG 11] OMG, "Business process model and notation (BPMN)", v2.0, available at: http://www.bpmn.org/, 2011.

[PAU 93] PAULK M.C., CURTIS B., CHRISSIS M.B. *et al.*, "Capability Maturity Model for Software, Version 1.1", CMU/SEI-93-TR-24, ESC-TR-93-177, Software Engineering Institute, Pittsburgh, 1993.

[TOL 07] TOLK A., DIALLO S.Y., TURNITSA C.D., "Applying the levels of conceptual interoperability model in support of integratability, interoperability, and composability for system-of-systems engineering", *Systemics, Cybernetics and Informatics*, vol. 5, no. 5, pp. 65–74, 2007.

[VAN 12] VAN STADEN S., MBALE J., "The information systems interoperability maturity model (ISIMM): towards standardizing technical interoperability and assessment within government", *International Journal of Information Engineering and Electronic Business*, vol. 5, pp. 36–41, 2012.

[VER 99] VERMA D., *Supporting Service Level Agreements on IP Networks*, Macmillan Technical Publishing, London, 1999.

[YAH 12] YAHIA E., AUBRY A., PANETTO H., "Formal measures for semantic interoperability assessment in cooperative enterprise information systems", *Computers in Industry*, vol. 63, no. 5, pp. 443–457, 2012.

[ZAD 78] ZADEH L.A., "Fuzzy sets as a basis for a theory of possibility", *Fuzzy Sets and Systems*, vol. 1, pp. 3–28, 1978.

7

Interoperability and Supply Chain Management

7.1. Introduction

Over the past 20 years, the logic behind the activity of organizations evolved from being focused on *products*, in the broadest understanding of the term, to being focused on the sustained satisfaction of customer needs. For these organizations, the notion of performance has shifted from being oriented towards the traditional triptych "costs, quality, timing" to a much more global understanding. Performing well on these three dimensions is nowadays necessary but not sufficient. Many other factors must now be taken into account to ensure the competitiveness of an organization and, more importantly, its perpetuation. Dimensions such as *reactivity*, *flexibility*, *robustness*, *stability* or even *resilience* have become key to the determination of the performance of a company and, *a fortiori*, of a supply chain (Figure 7.1). For managers, the challenge nowadays consists of learning to deal with the contingencies and perturbations that have become inherent to the daily life of companies.

The supply chain concept has long been a matter of debate. Supply chains are now commonly understood as networks of organizations – supporting physical, informational and financial flows. The networked organizations are involved in upstream and downstream relationships around various processes and activities and provide a product or service with the aim of satisfying the customer [CHR 92]. This relatively consensual definition nevertheless stresses the complexity of such networks. Mentzer *et al.* [MEN 01] note that distinguishing supply chains from their management (*Supply Chain Management*) is fundamental. Supply chain

---

Chapter written by Matthieu LAURAS, Sébastien TRUPTIL, Aurélie CHARLES, Yacine OUZROUT and Jacques LAMOTHE.

management requires that all agents involved in value creation engage in voluntary action, whereas supply chains merely define an existing environment in which processes are executed. Over the past 20 years, professionals in the logistics field have invested a significant amount of effort into setting up a global and coordinated stewardship of their networks in order to provide better customer service whilst limiting costs.

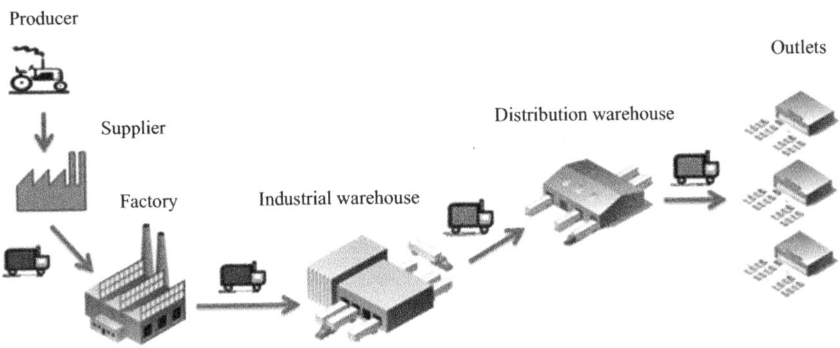

Figure 7.1. *Example of supply chain*

On a practical level, we find two types of collaboration: transactional collaboration and informational collaboration.

Transactional collaboration mainly seeks to increase the efficiency of business transactions between companies. It involves:

– transmission of normalized documents via electronic means (orders, dispatch notices, invoices, etc.);

– pre-invoices (in the supplier's name);

– availability of electronic catalogs (pre-referencing, portals), etc.

In this context, one of the most commonly used technologies is electronic data interchange (EDI), a framework for the transmission of dematerialized documents. This is the first step on the path to collaboration. There nevertheless still exists a "border" between the two involved parties, because we do not have any visibility on the internal processes of the other. They only interchange transaction-related information.

Informational collaboration consists of putting into practice tools and methods that enable exchanges of information such as:

– demand or stock levels (activity forecasting, real end-customer sales – sell-in/sell-out);

– information on planned promotional offers;

– knowledge of supplier's production capacity, etc.

To develop this type of collaboration, we can apply several tools and methods such as VMI (Vendor Managed Inventory), DRP (Distribution Requirements Planning), CPFR (Collaborative Planning, Forecasting and Replenishment), etc.

The remainder of this chapter aims to show how interoperability contributes to fulfilling the needs for transactional and/or informational collaboration in modern supply chains.

7.2. Supply chains interoperability needs

The evolution of supply chains can now be seen as a rigid structure transforming into a fluid environment. This metaphor highlights the importance and fundamentally changing nature of inter-company collaborations. Collaborations are no longer based on long-term perspectives involving stable relationships, the setting up of which is long and demanding. They are now based on reactive, one-time collaborations that can be quickly set up and dissolved.

Since supply chain agents want to quickly be able to work together whilst maintaining a definite level of confidentiality, information access must be controlled. This necessary limitation of partners' access severely constrains the collaboration in terms of data format and interchange coordination in heterogeneous technological and organizational environments.

These new demands explain why supply chains have inescapable needs in terms of interoperability. The important take-home message is that interoperability is the necessary solution to the information interchange needs of heterogeneous organizations. This interchange should ideally be implementable without any of the partners needing to make a specific effort, i.e. without each new collaboration requiring specific modifications.

Supply chain agents are, in essence, independent from each other. This notion of independence leads supply chains to favor interoperability over integration (see Table 7.1).

Interoperability	Integration
Maintained identities and diversity	Assimilation
Loose coupling (independent components)	Tight coupling (interdependent components)
No normalization of vocabulary	Normalization of vocabulary
Correspondence of global objectives	Correspondence of global and local objectives

Table 7.1. *Difference between interoperability and integration*

Thanks to advances in information science, information, hardware and software architectures have become the main means to satisfy interoperability needs within supply chains. The Information System impels structures to open to each other. Reducing the issues of supply chain interoperability to an Information System issue would however be exaggerated. Logistic relations always grow on the basis of existing activities and on the will of involved parties.

7.3. Various types of supply chain interoperability

Supply chain management implies transversal interoperability across several interdependent entities on the legal, institutional and strategic planes, without these entities establishing any subordination relationship. In order for the supply chain to meet its structural goal (product and service delivery to the end-customer), each entity must complete a mission for which a given performance target has been defined previously. To achieve this, each entity uses its own action or decision variables, taking into account constraints and pre-defined criteria. Vallespir and Doumeingts [VAL 02] call such entities "decision centers". Any supply chain can be analyzed into a chain of supplier–customer binary relations. Let us here consider a system consisting of two decision centers that share a common goal on a given process. The notion of the decision center is eventually defined as a set of activities all oriented towards the same planning horizon and the same planning re-evaluation schedule. Activities must be performed according to shared goals defined by a single, shared, decision framework [BRE 84]. This decision framework typically consists of a set of input and output relations, liaison variables, internal or external constraints and, of course, goals [ROB 93].

In the field of supply chains, various notions can be invoked to initiate or advocate for partnerships identified as necessary. We speak of communication, cooperation, collaboration, coordination and even of decision center partnerships. The object of an inter- or intra-company relationship is hardly more precise than the setting up of a kind of association. This definition does not explicate the diversity of actual exchanges and sharing procedures that can be put to work. For example, nothing in an associative relationship description enables us to differentiate the intrinsic features of an ordinary supplier/customer relation from those of purchaser/subcontractor relationships, or from those of a producer/distributor relationship. Furthermore, companies not only face the dilemma of “doing” or “making do”, which they resolve by a costs/benefits arbitration, but also face constraints due to unpredictable events arising from the specific features of the processes they put in practice and have to manage. So, the question is: why do various interchange and sharing modes coexist? In order to understand it, we must investigate the notion of decision center *interoperability*, defined by Lauras *et al.*, [LAU 03] as:

– an apparatus by which decision centers partner up (definition of a common goal in a win–win relationship);

– the taking into account of unexpected events;

– a collective joint action that overcomes individual limitations.

The various states of interoperability between decision centers can be determined in relation to, on the one hand, temporal aspects (one-time interchange, formalized interchange process) and, on the other hand, the objects they operate and the mode of transmission they use (exchange of type [A output = B input]/sharing of type [A and B put the same information to different uses]). Decision centers handle objects that can be categorized into two distinct types: data and processing. The term “data” is used in a broad sense. It encompasses the notion of information as well as that of parameters provided as inputs or outputs of a decision center. The provided “data” is not contextualized by the decision-making process. The term “processing” encompasses operating modes, decision variables and constraints to be taken into account when determining the manner in which decisions are made within a decision center, independently of the timed context determined by the “data”.

We would like to remind the reader that our aim here is to identify the various types of interoperability that exist between two decision centers involved in a collaborative process, and more specifically in a logistic process. The name given to any state of the understanding between the partners is not the essential element; it mainly serves as a semantic vector of the determinations it conveys. We, however, propose to define the various elementary states of interoperability (see Figure 7.2) between two decision centers A and B, according to the transmitted information (data or processing) and its mode of transmission (one-time exchange, sustained interchange or sharing process).

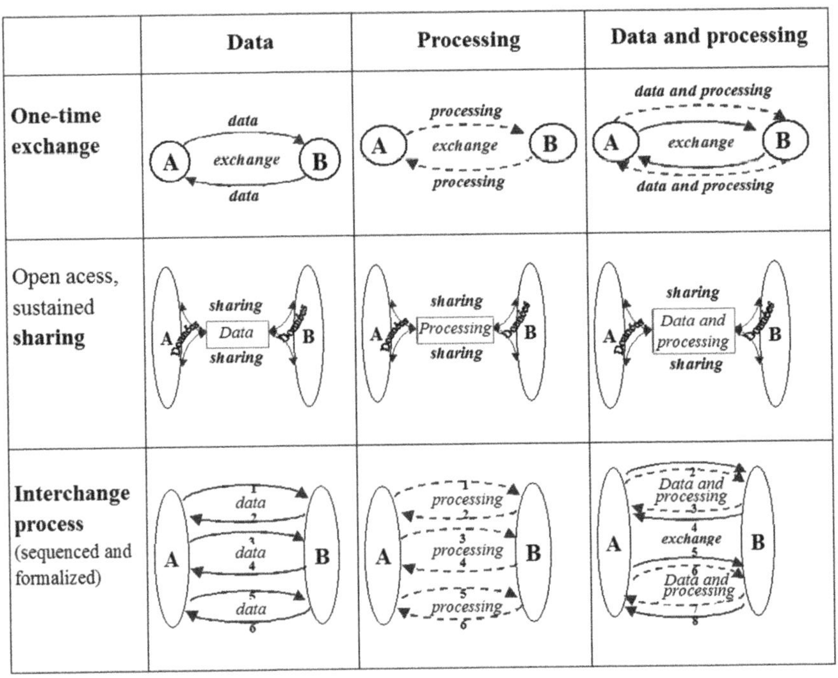

Figure 7.2. *Characterization of the elementary dimensions of interoperability*

These states of interoperability are considered elementary because associating them enables the characterization of the various states of interoperability that we encounter, for example, in a supply chain. To ascribe names to the various states of interoperability, we restricted ourselves to the following terms: communication, coordination, collaboration and partnership. We associated these terms with elementary characterizations of the various types of interoperability. The positioning we present here corresponds to a consensus reached by several studies[1] on supply chains. A relative consensus emerged about the notions encompassed by the terms "communication" and "coordination". According to this frame of reference, types of interoperability are characterized by these notions in the following manner:

– a one-time data exchange corresponds to "communication". To provide information about delayed dispatches or ordering mistakes corresponds to an example of logistics communication between decision centers;

– a sustained, regular and structured understanding established either by a formalized data interchange system or by a data sharing scheme corresponds to "coordination". The creation of a database of, for example, product references shared by the two decision centers is a good illustration of coordination;

– "collaboration" and "cooperation" are however often considered as quasi-interchangeable synonyms.

We chose to limit the domain covered by the term "collaboration" to:

– a one-time processing or data and processing interchange. The determination of goals and procurement frequencies by mutual agreement between the two decision centers is an example of one-time processing interchange;

– processing interchange.

Conversely, the term "cooperation" encompasses several notions, whose complexity varies according to how elementary states are combined to characterize the desired state of understanding. In that case, cooperation is understood as a regular and sustained long-time relationship. Several features are associated with it:

– formalization of the processing interchange procedures;

– formalization of the data and processing interchange procedures;

– data and processing sharing;

– combination of the three previous configurations.

"Partnership" is seen as the more or less formal contract agreement of one of the interoperability states (elementary states or combined elementary states). The generic quality of the term is therefore maintained, as it can apply to any interoperability state.

1 PhD dissertations of Lauras [LAU 03], Parrod [PAR 05] and Telle [TEL 03].

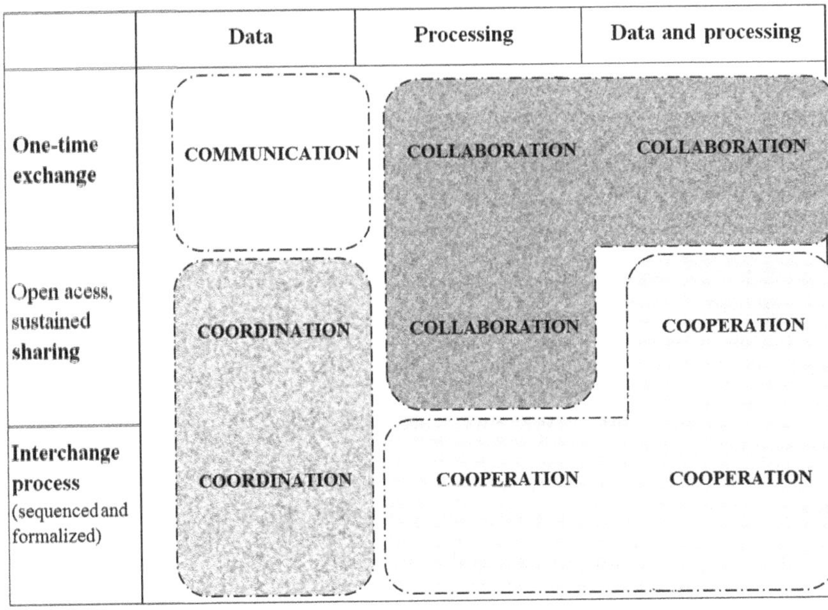

Figure 7.3. *Vocabulary of the characterization of supply chain interoperability*

The various types of supply chain interoperability have been defined. We shall now discuss the application vectors of these types of interoperability: Information Systems.

7.4. The main logistic Information Systems to support interoperability

The level of maturity of a company in the field of logistics directly influences its choice of Information System. If the various departments of the company are strongly compartmentalized and if there is no network of collaboration with, for example, suppliers and/or clients, then the first thing to do is to set up basic tools to computerize and share at least parts of the data. In that case, the work mostly involves setting up project management aids, such as ERP systems. Once at least one part of the information is computerized, we can move on to setting up tools to ease data interchange between partners (see Figure 7.4).

	M/V Operations Traditional logistics	M/V Operations Functional supply chain	M/V Operations Internal supply chain	M 0 Extended supply chain
Focus	Cost, quality, deadline	Customer service	Lower cost customer service	Profitable growth
Organization	Independent departments in silos	Improved operation	Supply chain department	Collaborative network
Process	Standard operating procedures	Interfaced processes	Transverse processes	Collaborative processes
Performance	Foreseeable costs	Complete and timely delivery	Transversal indicators, total cost after delivery	Shared indicators
Decision aids	Paper	Spreadsheets	APS	CPFR
Execution aids	MRP & specific applications	MRP II and specific software (ARP)	ARP WMS/TMS/MES	Web server

Figure 7.4. *Influence of logistic maturity on the choice of Information System*[2]

Several tools to support operational management of supply chains exist and the necessary interoperability can be based on any one of them. Tools are usually categorized according to two criteria: their functional purpose in the company or network of companies and the time perspective of their application (see Figure 7.5).

In a long-term perspective, APSs (advanced planning systems) enable the simulation and planning of the logistic network. Such software packages optimize planning and synchronize the flows of the supply chain, simultaneously taking into account many constraints (resources, capacities, delays, costs, benefits, etc.). APSs are more recent than ERP systems. They were initially aimed at specific business sectors, in which they tackled specific issues (such as the design of logistic networks). Their functional and technological scope is currently being extended, so that they can be adapted to almost any decision-making situation of a strategic nature. They have thus become decision support tools that help globally optimize most of the company's functions (procurement, production, transport, distribution, etc.). They can be based on linear programming techniques or on genetic algorithms. They require fine-grained knowledge of the whole of the operationally managed system.

Mid-term, companies use ERP (Enterprise Resource Planning) systems, which cover most of the diverse functions of the company, from procurement to human

2 UNITEX and Cap Gemini, "Comparative study of global supply chains", *Conference on the Technical Textile Industry*, Lyon, Rhône Alpes, France, 2005.

resources. These tools gather and integrate within a single database all the data and managing knowledge of the company. An ERP is often the combination of several application modules (one module per regular analytical function of the company). The modules can communicate within an interchange framework because they are all based on a single database, which means that the processed data is uniform. Each module therefore receives information from other modules and, in turn, distributes its own data to other modules. ERP systems are a very popular choice of supply chain management tools. They have nowadays become a true standard of organization-oriented Information Systems, especially in the field of supply chain management. However, the benefits enabled by ERP systems should not make us overlook their complexity, as companies sometimes end up considering them as the cause of all their woes [MOS 10].

Short-term tools include, for example, MESs (Manufacturing Execution Systems) and SCE (Supply Chain Execution) systems, which include WMSs (Warehouse Management Systems) and TMSs (Transport Management Systems). A WMS is a software suite to optimize warehouse management. It comes with several functions that ease reception, storage, preparation, dispatching, identification and operational management of products. A TMS offers a logistic solution to enable planning, freighting, physical execution management, operational records, quality monitoring and cost control. A TMS also enables the taking into account of environmental issues. An MES is a software package that provides real-time significant information about execution of manufacturing orders. It enables the monitoring of their progress from the initiation of production to the completion of products, as well as optimization of the production activity.

Setting up an Information System may challenge several settled habits and customs of the company and its network of partners. The integration of an Information System may disrupt the smooth running of existing processes and procedures and therefore deteriorate some performance indicators. Let us quote, for example, Richard Wolff, director of the SAP board of Sopra Group in an interview of Journal du Net Solutions[3] about logistic interoperability projects: "*One sometimes makes the mistake of imagining that this type of deployment project is a pure computer science issue. This is an error because such projects impact the processes of the company and therefore influence the daily experience of users*". We believe that the issue goes beyond company processes, as such projects also impact inter-enterprise processes!

3 Ludovic Tichit, *Réussir son ERP.* (Successful ERP deployment), JDN Solutions, July 2006.

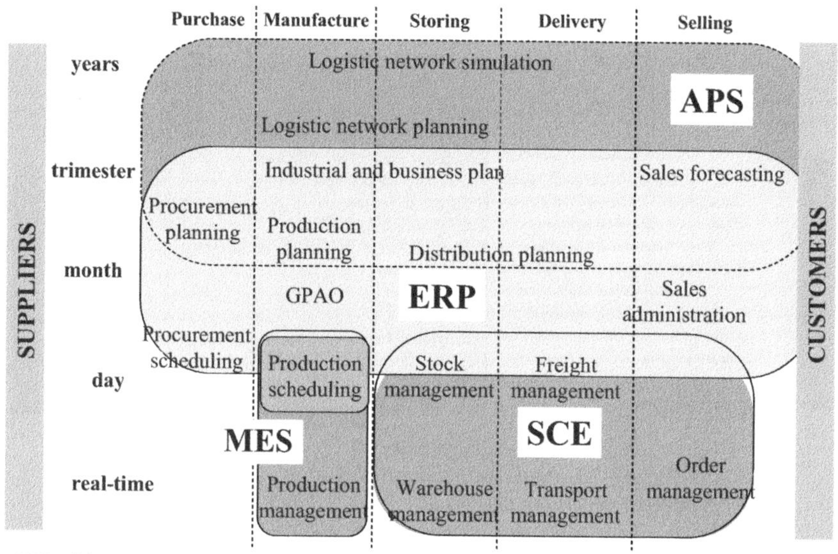

APS = Advanced Planning System

ERP = Enterprise Ressource Planning

SCE = Supply Chain Execution

Figure 7.5. *Classification of the main available tools*

Change management is therefore an essential dimension of logistic interoperability projects. Change management involves more than training courses concerning the new system. A significant fraction (about 15%) of the budget must be invested in change management in order to successfully deploy ERP systems.

CSC (Computer Sciences Corporation) and the journal Supply Chain Management Review (SCMR) conducted a worldwide study in 2008. They interviewed logistics professionals of 22 business sectors (production, distribution, warehousing, etc.). This study analyzes advances in the field of logistics as well as issues that have emerged; 134 companies participated. Of the total replies received, 32.7% came from outside North America. Europe is fairly significantly represented, with replies from 10 European countries, including 12 replies from France. The study highlights several results. One of them concerns the relation between the introduction of new technologies and the improvement of the overall performance of the company: 34% of all replies considered that the relation is "very efficient" or "quite efficient", while 31% considered that it is less definite.

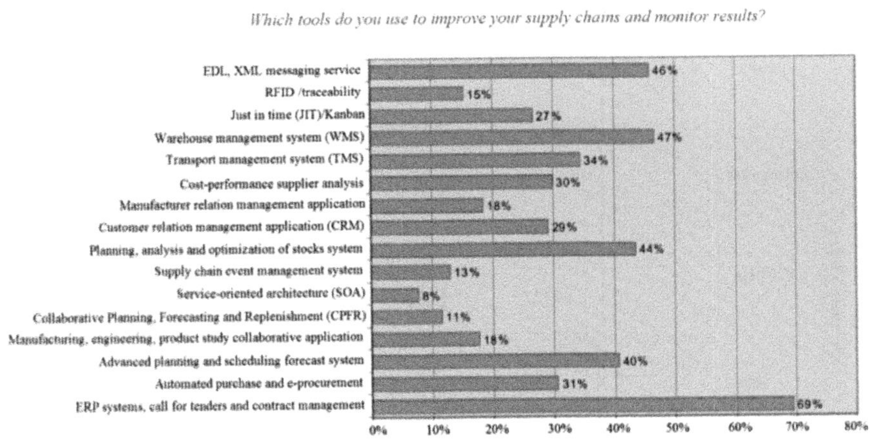

Figure 7.6. *Main Information Systems used in the context of supply chains and their relative importance*

In order to achieve a satisfying level of service performance, all Information Systems and processes that affect customers (order entry, order fulfillment and invoicing) must operate consistently and flawlessly. Being able to integrate one's internal systems and processes with those of important customers is a major asset. This is another essential aspect of interoperability needs, which does not seem fully covered by the current state-of-the-art offer.

Let us illustrate this point with the case of Shell's logistic network. Shell's automatic procurement system ensures that customers do not need to enter orders anymore, never face stock shortages and do not need a safety reserve anymore. On the contrary, their systems interoperate with Shell's systems in order to exchange real-time information that helps the operational management of flows. Consumption data, stock levels and demand forecasts are recorded daily on the customer's website. They are then communicated to Shell. When figures reach a predefined threshold, Shell generates a customer procurement order, prepares the dispatch voucher and monitors transport all the way to delivery. The system operates without any invoicing. Each month, the customer electronically pays an amount evaluated on the basis of the consumption figures exchanged by the two companies. This solution drastically lessens the customer's administrative costs as well as its safety reserves, thus lessening the total volume of stocks in the supply chain, which greatly simplifies its management.

Technology may be necessary to improve supply chain interoperability, but the setting up and implementation phases still remain significant causes of concern for companies. The deployment of a new logistic Information System must be considered as a true company-wide project and be prepared at all levels, from management to end-users. The stakes involved must be well explained. Key users play an important role in this process. In order for the transfer of competencies to be successful at the technical level, a good practice involves having the integrator train these key users. They will then be able to forward the most important information to their colleagues more efficiently. Furthermore, at the human level, they will be more suited to act as relays and motivate end-users in the field. Therefore, in order to ensure the successful completion of an interoperability project, we should proceed through several essential phases:

– initialization: this phase consists of defining the project (relating business needs to how the Information System fulfills them), drawing up a functional specification document and choosing one solution;

– target system design: this phase consists of modeling internal and collaborative business processes (gap analysis), determining technical architectures and logistic dimensions as well as analyzing costs;

– implementation: this phase consists of a detailed parameterization of the solution, followed by the migration from the previous solution to the new one;

– optimization: this phase consists of monitoring indicators, stabilizing, fine-tuning and optimizing the solution as well as developing necessary additional applications and extensions;

– information and communication: This phase consists of informing and communicating, both internally and externally, about project advancement and evaluation.

7.5. Main architectures to support logistic interoperability

Classically, we find two types of architecture to support supply chain logistic interoperability: the first one is a “point-to-point” linking enabling the “hard” linkage of the various Information Systems involved in supply chain management processes. The following figure shows this type of architecture.

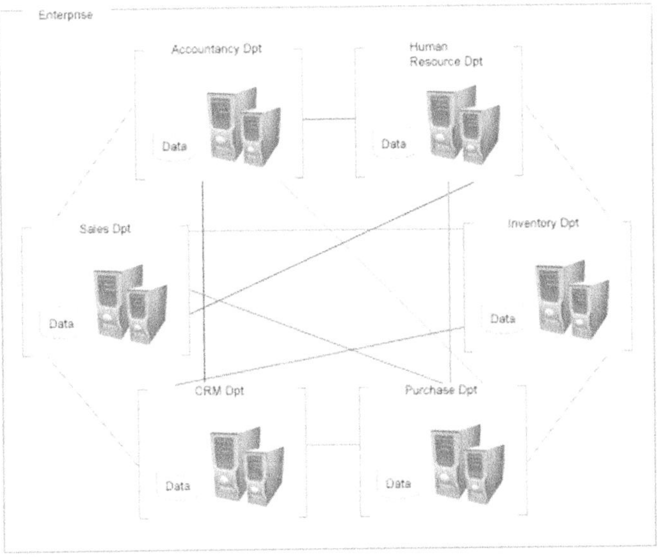

Figure 7.7. *Architecture of the point-to-point type*

The second type of architecture emerged over the past few years with architectures called *Service Oriented Architectures (SOA)*. The SOA denotes a style of architecture composed of several distinct services attached to the various activities involved in company processes. Architectures of the SOA kind can be used to relate the activities of the company to application services provided by the Information System. This relation is the key element to relate a collaborative process, which is defined at the activity level, to a "technical" process to be executed by one or several Information Systems. Services of SOA architecture-based Information Systems are characterized by access points. Access points constitute the public side of the service. Each access point comes with a service-level agreement. Such a service-level agreement specifies both the inputs accepted by the service (via the corresponding access point) and the structure of its output messages.

Such service-level agreements are directly inspired by professional business service-level agreements. Service-level agreements are documents that detail all the necessary elements of description of the service. In the business relation world, a service-level agreement is a document that has several parts. Its terminal elements are usually called articles and clauses. In the SOA world, service-level agreement elements specify service functions, exchanged messages, communication protocols and the physical address of the service.

In order for the collaboration to run smoothly, it is important that a service-level agreement be both readable by humans and exploitable by software. To this end, service description languages have been developed. The most commonly used is WSDL (*Web Services Definition Language*). Each service comes with a WSDL document that specifies the service-level agreement.

SAO architectures are usually organized around an *Enterprise Service Bus (ESB)*. ESBs, as their name suggests, are based on the concept of "bus". Service distribution is one of the main advantages of such an infrastructure [TOU 07]. ESBs enable indirect connections between various services of Information Systems: once a service provided by the Information System is connected to the bus, it is potentially linked to all the already connected services, even if these services are based on different technologies. This is illustrated in the following figure.

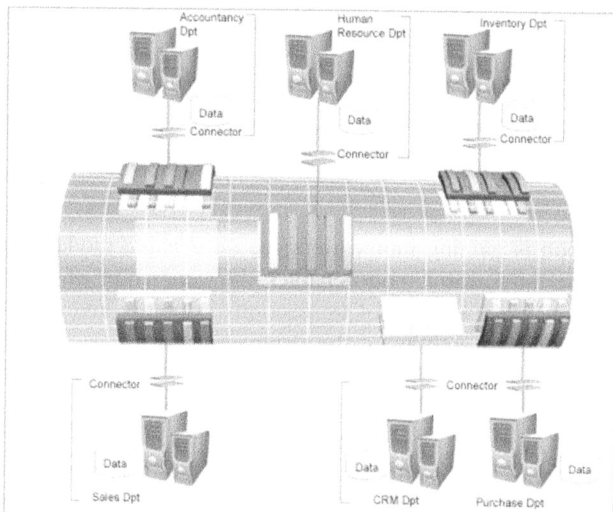

Figure 7.8. *ESB-type architecture*

7.6. SaaS applications revolutionize logistic interoperability

Within the SOA trend, supply chain management-dedicated software applications have, over the past few years, undergone a true revolution due to the appearance of Software as a Service (SaaS). This section aims to provide an overview of the extent to which this type of software has disrupted the habits of supply chains and to highlight the resulting benefits in terms of interoperability.

Often, companies choose to implement their logistic interoperability by way of purchasing business-specific ad hoc software solutions (licenses) and connecting them to each other (see the point-to-point architecture we detailed above). Such software licenses are invoiced as a function of the number of users. Data is hosted on a local server. The company is therefore in charge of ensuring the safe storing of its own data. Service-level agreements (maintenance and support) are contracted with the software editor and the amounts involved are often higher than licensing costs.

Other companies choose the IT outsourcing mode. According to the AFNOR (French Standardization Association), IT outsourcing is defined as the result of the integration of several elementary services into a package in order for a customer to entrust parts or the whole of its Information System to an IT service provider, with a flat-rate agreement specifying the level of service and its duration that may span several years. We thus speak of IT outsourcing when the software editor provides the company with a single contact entity who will be in charge of the project throughout its whole life cycle, from the purchase of the software to its maintenance and exploitation. The information architecture remains the property of the company and interoperability is limited to its lower level of integration (point-to-point fixed connections).

Another solution is gaining momentum: the *on-demand* mode. In contrast to a yearly license, the *on-demand* mode enables the generation of monthly detailed invoices that list the exact number of transactions. Monthly invoices with amounts proportional to the actual activity of the company have been commonplace for a long time, with a simple method consisting of simply updating the number of licenses according to activity peaks. The difference between that trick and the current *on-demand* systems lies at the technological level. Current *on-demand* systems leverage the latest technological advances, the most well known of which being SaaS (Software as a Service). Data is hosted on an external server, which does not belong to the company. The software editor is in charge of this server. Remote access is usually provided via the Internet. Smartphone applications have lately been developed in order to make access even easier.

According to Didier Santurette, from Acsep, "*On demand* applications have emerged from the will of customers to increasingly tightly adjust their costs"[4]. This observation applies to companies of any size. It however especially applies to SMBs, who often have difficulty heavily investing in hardware and thus favor this lighter solution for its lower costs.

4 *Le journal de la logistique*, November 2011.

SaaS-mode solutions are also often chosen for reactivity and simplicity reasons. Large groups such as Total or Airbus Helicopter have, for example, opted for the SaaS mode under this rationale, even if it required them to redesign their warehouse management procedures to adapt them to the new software solution.

The software editor must then be able to quickly adjust its available computing power to customer company demands. The customer company is not in charge of maintenance or hardware renewal costs anymore.

Furthermore, in the SaaS mode, the data storage platform is shared and related costs are therefore curbed.

Last but not least, it applies to all the tools listed in this addition: in the SaaS mode, software updates and upgrades are managed by the software editor. Each customer company is therefore guaranteed to use the most up-to-date version of the software.

Focusing on specific logistic services will now enable us to list more benefits of that approach. SaaS are indeed obviously advantageous when using TMS and WMS.

In the case of TMSs, detailed invoices listing each transaction are undoubtedly beneficial. Depending on the business contracts they manage to sign, TMS partners may experience significantly variable activity. Their internal Information System skills might furthermore be limited.

Another advantage of using a TMS is the capacity it grants a collaborative platform or company to distribute the TMS-related costs among all involved transporters.

Moreover, since TMSs connect a great number of agents, SaaS-type procedures enable all involved parties to easily share information. The SaaS mode also enables companies to manage their transport issues without having to also concentrate on Information System issues and constraints, which can be especially complex in the context of security management, an essential aspect of collaborative environments.

Although the numerous benefits brought by SaaS convinced many companies to adopt this mode of information management, we should not overlook several drawbacks involved with using the SaaS mode.

Most SaaS solutions are very generic and leave very little room to adjust the proposed solution to specific company needs. Specific needs then become add-on applications that customer companies need to buy in addition to the core solution, following the same business model as, for example, iPhone apps. Implied by this extension model is the requirement that each extension be developed by the software editor. According to SAP, the specific requests of companies are processed when the number of requests reaches a predefined threshold or when the idea seems potentially interesting for several other customers, which will ensure the profitability of the application.

Most companies also have qualms when they realize they must entrust their security and the secrecy of their data to a third party.

Lastly, many companies have already significantly invested in hardware to support other Information Systems. In that case, adding an application does not incur heavy investments.

According to Tom Eid, research vice-president of Gartner, "After more than 10 years of use, SaaS technologies keep increasing their share of the market of enterprise information applications. This growth takes place in a context where ever-tighter budgets demand less costly solutions. The popularity and fame of the SaaS model increase in parallel with the interest for *PaaS (Platform as a Service)* systems and *Cloud Computing*".

Tom Eid then adds, "Initial concerns regarding security, response delays and service availability gradually abate in many organizations as the SaaS market and computing models mature whilst usage is increasingly generalized. On-demand solution provider ecosystems and usage keep evolving towards more diverse technological and management services, more verticalized functions and stronger partner and purchaser communities"[5].

As mentioned earlier, most TMS and WMS software editors keep developing solutions for medium- and large-sized companies, although almost all of them offer at least one SaaS-mode solution.

Many ERP solutions have recently been released, most of them oriented towards SMBs.

---

5 http://www.erp-infos.com/info_article/m/1427/21--de-croissance-du-marche-mondial-du-saas-selon-le-gartner.html

7.7. Conclusion

Nowadays, supply chain management and, more broadly, inter-enterprise collaboration can hardly be discussed without considering interoperability issues. Owing to recent advances in information science, information, hardware and software architectures have become the main means to implement interoperability. To understand the whys and wherefores of this technological domain has thus become a necessity for all supply chain agents. According to its size and business sector, a given company needs to carefully select its software applications and corresponding architectures. Most importantly, companies must think beyond Information System issues and fully understand business relations woven across the various agents of the supply chain. Finally, interoperability needs will vary according to the level of maturity of the collaboration.

7.8. Bibliography

[BRE 84] BREUIL A.D., Outils de conception et de décision dans les organisations de gestion de production, PhD Thesis, Toulouse 1 University of Social Sciences, 1984.

[CHE 08] CHEN D., DOUMEINGTS G., VERNADAT F., "Architectures for enterprise integration and interoperability: Past, present and future", *Computers in Industry*, vol. 59, no. 7, pp. 647–659, 2008.

[CHR 92] CHRISTOPHER M., *Logistics and Supply Chain Management*, Pitman Publishing, London, 1992.

[LAU 03] LAURAS M., PARROD N., TELLE O., "Proposition de référentiel pour la notion d'entente industrielle : trois approches dans le domaine de la gestion des chaînes logistique", *Revue Française de Gestion Industrielle*, vol. 22, no. 4, pp. 5–30, 2003.

[LAU 04] LAURAS M., Méthodes de diagnostic et d'évaluation de performance pour la gestion de chaînes logistiques: application à la coopération maison-mère–filiales internationales dans un groupe pharmaceutique et cosmétique, PhD Thesis, National Polytechnic Institute of Toulouse, 2004.

[MEN 01] MENTZER J.T., FLINT D.J., HULT G.T.M., "Logistics service quality as a segment-customized process", *Journal of Marketing*, vol. 65, no. 4, pp. 82–104, 2001.

[MOS 10] MOSCAROLA, *et al.*, "Usage d'un ERP (Enterprise Resource Planning) et processus de gestion: cas d'une entreprise agroalimentaire", *La Revue des Sciences de Gestion*, no. 243–244, pp. 143–151, 2010.

[PAR 05] PARROD N., Analyse d'un processus de coopération entre donneur d'ordres et sous-traitant au sein d'une chaîne logistique projet: une approche par simulation, National Polytechnic Institute of Toulouse, France, 2005.

[ROB 93] ROBOAM M., *La méthode GRAI. Principes, outils, démarche et pratique*, Teknea 1993.

[TEL 03] TELLE O., Gestion de chaînes logistiques dans le domaine aéronautique: aide à la coopération au sein d'une relation donneur d'ordres-fournisseur, PhD Thesis, 2003.

[TOU 07] TOUZI J., Aide à la conception de Système d'Information Collaboratif, support de l'interopérabilité des entreprises, PhD Thesis, National Polytechnic Institute of Toulouse, 2007.

[VAL 02] VALLESPIR B., DOUMEINGTS G., "La méthode GRAI. Groupement de Recherche en Productique (GRP)", *Ecole de printemps, la modélisation d'entreprise, Une réflexion sur l'enseignement et la pratique des méthodes*, Albi, 28–30, 2002.

8

Organizational Interoperability Between Public and Private Actors in an Extended Administration

This chapter discusses organizational interoperability issues between public and private organizations in an extended administration. It presents a framework that can help to manage this interoperability and to evaluate the network performance, by identifying the specific interoperability requirements for a public–private network.

8.1. Introduction

Public administrations face two challenges. The first one is to define the needs they must satisfy. This issue occurs for needs of the users of public services and also for the needs of the civil servants (so they can perform their tasks). Although it is a challenge that all organizations face, in the case of the public sector the definition of the needs involves taking into account potentially conflicting mechanisms: market mechanisms and public interest. The second challenge is to implement the organization that is capable of satisfying these needs. These two challenges are closely related. Changing needs may affect the organization and vice versa. Administrations must then cope with continuous adjustments.

When the needs are standard and stable, i.e. common to all and slowly evolving, administrations face few difficulties in their organization. Standardized and stable

Chapter written by Yacine BOUALLOUCHE, Raphaël CHENOUARD, Catherine DA CUNHA and Alain BERNARD.

needs are conducive to a Weberian administration [WEB 63]: a centralized and hierarchical organization. Yet the current analysis stresses the existence of personalized and evolving needs. The ongoing re-engineering of administrations in order to adapt to these needs showed the limits of the Weberian model: a lack of relevance of the offered services and a lack of parsimony in resource management [LOR 99]. Thus, from the 60s emerged a "self-preoccupation of the State" [BEZ 09] highlighting the need for the state and the administration to reorganize.

New Public Management (NPM) has inspired the majority of the reforms in the public administration over the last 30 years [OSB 93]. Hood was the first to define NPM as an answer to the heaviness of the Weberian administration [HOO 91]. NPM aims to redefine ways of action and at the same time the perimeter of administrations' intervention. It is based on the ideology that the use of management methods traditionally applied in the private sector leads public organizations to more efficient use of their resources and the achievement of their goals. But the NPM has caused fragmentation of the public value chain, leaving a legacy of multiple public and private organizations engaged in public services [LEV 12]. Organizational difficulty has evolved and several countries have begun to amend reforms of modernization inspired by NPM. This evolution requires the establishment and management of networks to deliver public services and questions the interoperability of public and private organizations that compose them.

Administration networks multiply decision-making centers. Each center is a public or private organization with the skills to create others. This multiplication of decision centers makes it difficult to maintain the global coherence of public action. This complexity is increased by heterogeneity of organizations. This issue, mainly raised in the field of public management, can benefit from a new perspective and contributions from our community. It is from this perspective that this work takes place. Our contribution in this chapter is to establish a management framework which can help to manage interoperability and evaluate network performance by identifying the specific interoperability requirements for a public–private network. The next section will introduce the public–private network. Section 8.3 deals with interoperability in this context. Then, in section 8.4, we present our proposed framework before applying it to a case.

8.2. Public–private network

The term public service has several meanings. It refers, on the one hand, to the activities of general interest exercised in a direct or indirect way by the public authorities, and on the other hand, to all organizations (public or private) in charge of the activities of general interest. The public services therefore consist of public and private organizations. These organizations are said to be in a network since they interact to enable service delivery.

Among the various definitions of the concept of network, we can quote one of the first given by [THO 86]: a network consists of at least two companies linked by relationships strong enough to maintain an intermediate way of organization between the market and the hierarchy. These relations are long-term and their stability requires a minimum number of agreements and a mutual trust between actors. [BOU 90] specifies that "the only fact that two exchange relations, A–B and B–C share the same element B is not sufficient to conclude that the two relationships form a network A, B, C. Thus, two exchange relations A–B and B–C will be called 'connected' to B to form the A–B–C network structure only when the exchange between A and B will somehow affect the exchange between B and C, and *vice versa*." A network is the place of interactions where the occurring events on a given relationship have an impact on other relationships. Actions and decisions of an organization within a network therefore depend on those of others. This interdependence raises questions about the interactions between the organizations that form a network. These interactions should be monitored and controlled, without depriving organizations of their autonomy.

A network composed only of public organizations raises these issues. The management of interactions is difficult because the network structures are by definition legally "split". A public network is then composed of public organizations autonomous in their management although controlled by the state and/or local authorities. The complexity arises when it comes to including private organizations in the network [ROC 10]. The inclusion of a private actor in a network is described as a public–private partnership. The notion of partnership here covers all forms of openings and close links between the public and private sectors. The public–private partnership is a "lasting partnership between legally separate entities pursuing private and public aims and objectives, which, taken separately, could not be achieved with efficiency and economy in the context of programs, projects or operations of general interest, common good or public service" [MAZ 08]. The public–private partnership therefore contractualizes agreement between the partners who are satisfied with the convergence of their interests. The public partner commissions an activity at an attractive cost, which would have demanded him specific expertise and/or necessary resources. The private partner receives the opportunity to benefit from an activity traditionally performed by the public partner [HAF 09].

Public–private networks have considerable importance due to the diversity of services they deliver, as well as from an economic point of view. For example, in France, the budget entrusted to private organizations by public organizations, through public procurement, represents 10% of global GDP [CCI 11]. However, these networks have pitfalls:

– the lack of methods, models and tools to evaluate *ex ante* the overall performance of the contemplated scenarios for public–private partnership. This explains the differences observed between *a priori* satisfaction and implementation of the partnership. It can lead to public opinion's mistrust when the decision is made to entrust public activities to a private organization [AUB 05, MEN 09, BEL 09];

– a focus on the legal aspect to the detriment of the managerial aspect of public–private partnership [MAZ 09]. This leads to a need for methods and tools in order to ensure the consistency of the partnership project as a whole, across the entire relationship life cycle, taking into account that "the private sphere and the public sphere are specific areas for the organization of goods and services production methods, their organizational configurations, coordination mechanisms, structures and operating processes are different." [DUB 09];

– networks multiply decision-making centers. Each center is a public or private organization with the skills to create others. The risk is a focus of each center on its local performance, to the detriment of the overall performance of the public–private network in which they operate. An organizational coherence must be maintained for the entire network, whose decision-making centers are characterized by their heterogeneity. This requires exercising management (which is yet to be defined) on the decision-making centers to preserve overall coherence [HAF 09].

We suppose that the first issue can benefit from approaches developed to model and simulate value chains in extended enterprise [MAU 07, MED 13, DAA 11], but in this chapter we will not detail this topic. We also suppose that these three issues are related, and will focus on the field of interoperability which seems to be an interesting axis of research to answer them.

8.3. Inter-organizational interoperability

We have seen that the organizational evolution in the public sector forces the creation and management of the public–private network to deliver public services. These networks are made up of relations between organizations that need to be considered individually but also collectively. The complex management of such groups makes it mandatory to understand and face the dynamic characteristic of the networks (the network composition may change), as well as the evolution of relations between organizations and the heterogeneity of the organizations (each organization inside that network has different work and coordination methodologies). This heterogeneous set raises organizational interoperability problems [VER 07] and makes us focus more precisely on the concept of interoperability of collaborative processes between these organizations. The issue is to consider each organization-to-organization relation in order to locally optimize the production of goods and/or services. This should be done considering the "bigger picture", the whole network, in order to maintain an organizational coherence [RAU 14].

These relations are materialized as collaborative processes. A process is a "sequential, spatial and hierarchical organization of activities that requires resources (or means) to deliver products (or outputs)" [LAB 04]. A process is collaborative when it involves more than one organization. A collaborative process can be defined as a process whose activities belong to different organizations [AUB 02]. The process performance requires maintaining the compatibility between heterogeneous organizations, their capacity to coordinate and interact in their operational phase, while respecting and preserving their autonomy and their capacity to put an end to collaborations [MAL 12]. Consequently, the harmonization of the interactions while controlling the dependences between organizations is needed to avoid large-scale internal changes.

Since its early definitions, interoperability was characterized by its ability to increase heterogeneous systems' capacity to deal with such complexities [MOA 12]. In the 90s, the first definitions were linked to the information technology domains [IEE 90, ISO 93]. Interoperability has since been a work field and was extended to other domains. In our context, we retain the definition proposed by [PIN 09] who defines interoperability as the "capacity of systems, natively strangers one from another, to interact in order to establish harmonious and finalized collective behaviors, without the necessity of profound changes in their individual structure or behavior." In order to reach this interoperability, organizations must limit their heterogeneity [BLA 07]. For this to be reached, organizations must treat three categories of problems: organizational, conceptual and technological [INT 07].

Interoperability was gradually structured, categorized and characterized to respond to organizations' needs. Interoperability frameworks were defined: IDEAS [CHE 08], EIF [EIF 03], INTEROP [INT 07], IF [ATH 07] and FEI [CHE 09]. To these frameworks was added a standardization effort [ISO 98, ISO 00, ISO 09, ISO 11]. This leads the way in deploying interoperability at different management levels (business, process, service and data), while taking into consideration the barriers relating to enterprise interoperability (conceptual, technological and organizational) and the approaches to overcome barriers (integrated, unified and federated).

Many works were developed to let organizations increase their interoperability capacities. These works affect the measurement of potential interoperability, the measurement of the compatibility, the measurement of the interoperability's performances and finally the early detection of the interoperability's problems. The measurement of the potential interoperability has an objective of evaluating the readiness and maturity of a system to overcome barriers to interoperability created by the collaboration between different partners. For a detailed review of maturity models, refer to [PAN 06]. The measurement of compatibility consists, for each interoperability barrier (conceptual, technological and organizational) and management level (business, process, service and data), of observing the compatibility level of each partner [DAC 07]. The works on evaluation of the interoperability's performance propose to consider interoperability as a performance that can be measured through a measurement of operational efficiency. [BLA 07]'s work in this field is also a first attempt to proactively detect interoperability problems. According to [MAL 12], the proactive detection of interoperability problems consists of making sure that the formulated interoperability requirements are verified on a model of collaborative process. This allows providing (if needed) improvements before the partners participate in the collaborative process. To define an interoperability requirements repository, the authors define four classes of requirements: compatibility requirements, interoperation, autonomy and reversibility. These four classes are part of the life cycle of a collaborative relationship between two organizations (Figure 8.1).

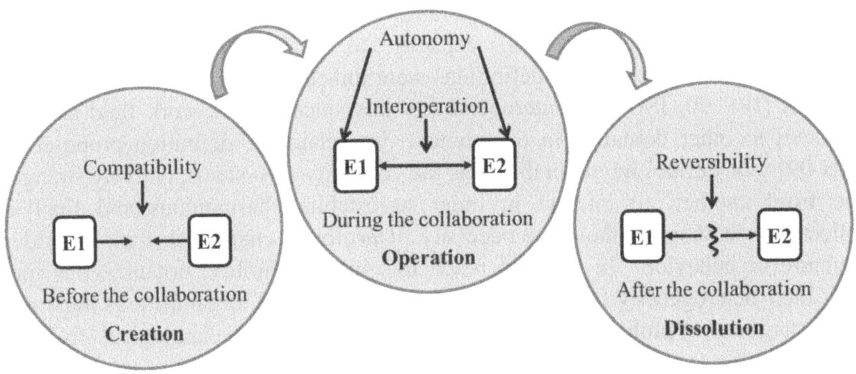

Figure 8.1. *Classes of interoperability requirements relating to the life cycle of the virtual enterprise [MAL 11]*

These works can offer the necessary methods and tools to manage collaborative processes in a public–private partnership. We can build on them in order to address the second issue (see section two). However, to the best of our knowledge, works taking into account the distinctive features of administrations and public services in order to maintain the organizational coherence of a network do not exist.

8.4. Management framework for extended administration

For public administrations, the challenge is to determine the public–private hybridization, developed on a network organization, enabling us to ensure the performance of their services. We call this mode of organization “extended administration”. Extended administration is the network organization of a service managed and controlled by public administration, made up of public and private organizations sharing the performance criteria defined by the administration. These performance criteria are a repository of assessment criteria, common to all organizations, of network activities. It has a scope: quantitative, as assessment criteria determine the choice of indicators to assess the performance of public–private networks, organizations and partnerships; cognitive, because it allows sharing a common vision of network performance objectives and thus to assess the relevance and consistency of the agreement between public and private organizations.

We identified three categories of problem with public–private networks:

– the public–private partnership performance evaluation;

– the partnership relationship management throughout its life cycle;

– the public–private network management as a whole, to ensure that local performance is not achieved at the expense of overall performance.

We suppose that these three categories are related: the public–private network management must be taken into account in relationships management and should guide, at the local level, the quest of performance based on overall objectives (Figure 8.2). It is a necessary condition for performance evaluation in a public–private network. The performance of an organization (public or private) as well as a partnership can then be evaluated on the basis of their ability to contribute to the network performance. Such an approach can contribute to the readability and legitimacy of the decision to use a private organization. However, this calls for considering the interoperability of management rules, models and performance indicators.

We suppose that this challenge can benefit from approaches developed to model, simulate and optimize value chains in extended enterprise. However, because of the distinctive features of administrations and public services, we observe limits on some of these approaches. Our contribution in this chapter is to establish a management framework for the extended administration. This framework must allow:

– the identification of specific interoperability requirements for a public–private network;

– the definition of the performance criteria;

– each entity to capture the specific requirements of the network, by adapting them to suit its compatibility, interoperation, autonomy and reversibility requirements regarding its direct partners in the network [MAL 12] and its internal processes [RAU 14].

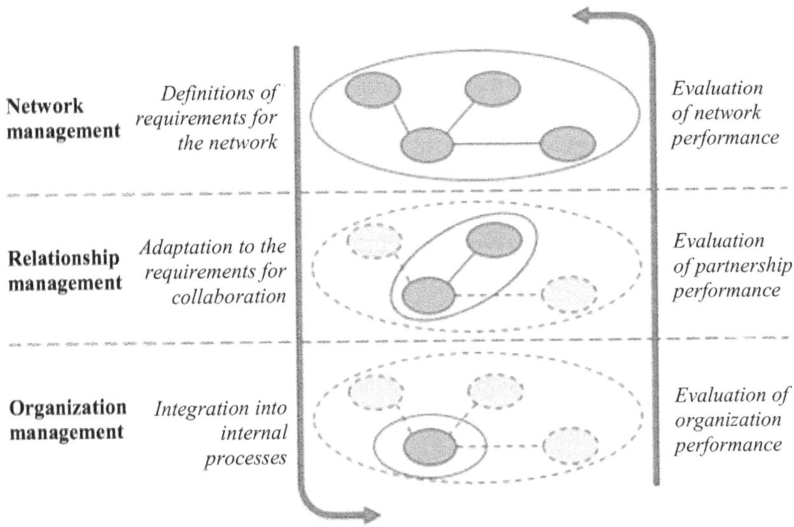

Figure 8.2. *Continuity of management levels*

A public–private network is a highly heterogeneous set composed of relations between heterogeneous public organizations and relationships between public and private organizations whose heterogeneity is accentuated by distinct culture. We must therefore define a management framework that applies to all and maintains an overall coherence [HAF 09]. From a state-of-the-art on public–private partnerships [AUB 04, AUB 05, BEL 09, DAV 09, MEN 09, MAZ 09] and a feedback from our industrial partner (CGI Business Consulting), we propose six areas of management to ensure organizational homogeneity for public–private networks (Figure 8.3):

– control of socio-economic impacts. We need to go beyond a short-term evaluation of financial performance by evaluating long-term socio-economic consequences of decision taken (avoid industrial imbalances, promoting access to SMEs, etc.);

– control of human resources aspects. There should be conditions of transfer and reintegration of employees (depending on their status);

– operational management. Assigned activities to a (public or private) delegate do not dispense the delegator to manage the proper implementation of activities;

– change management. This area allows benefiting from some flexibility in management activities and taking advantage of improvement opportunities;

– audit and control capacities preservation. We need to maintain public expertise and control activities;

– legal framework. Finally, we must define the legal framework that supports the application of other management areas.

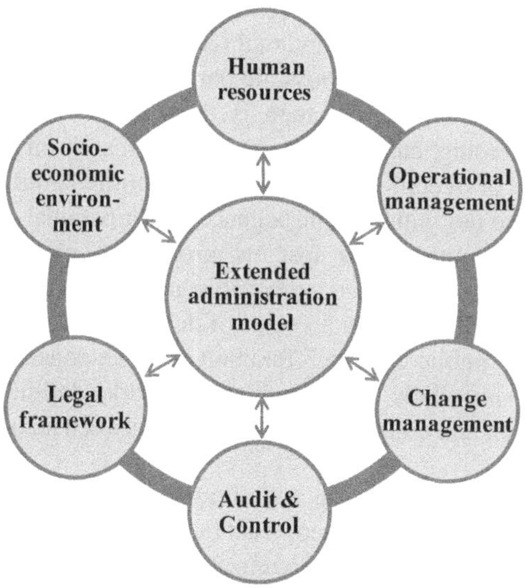

Figure 8.3. *Management framework for extended administration*

This framework aims at providing a holistic view of the administration's needs to maintain the organizational coherence of a public–private network.

8.5. Application to the "public clothing" function

A first application of this framework has been the definition of the performance criteria for the "public clothing" function. The textile and clothing industry often anticipates economic changes before many other industries [PER 08]. Its organization in extended enterprise around the world makes it an interesting experimental laboratory, while re-industrialization strategies are studied in France. The "public clothing" function composed of public and private organizations is, as a subset of this industry, an achievable research field. A function overview is performed by [OBS 12]. The "public clothing" function employs 2,750 workers to equip more than 1.6 million civil servants. It is a global market of about 650 million EUR. However, this market is fragmented as orders come from almost all public entities: the department of defense, national security forces (gendarmerie, police and firefighters), municipal security forces (police, firefighters), local authorities, SNCF, EDF, GDF, La Poste, etc.

Today, each of these entities organizes its "clothing department" as a network (composed for example by direction, internal teams, public and/or private operators) to meet their needs with the highest possible performance. For each of these "public clothing" submarkets, local performance is then sought. This leads to services rationalization, outsourcing, etc. We assume that the search for a local performance comes at the expense of overall performance for the entire "public clothing" market. Indeed, these steps are not without consequences for a national industry that, due to increased global competition, has lost its production capacity in the country [PER 08]. This raises questions about a profitable growth strategy for all stakeholders, and the "public clothing" market as a vector of development for the sector. For this, we propose to define "public clothing" function as an extended administration. This leads to rethinking the whole public–private network, defining the performance criteria for the function and collaborative processes between actors in the network.

Our first work is about modeling and simulation of business processes in this network. Generic activity models to represent business processes are proposed. They allow initiating dialogue between organizations with a common language to represent everyone's activities in the network. This notion of a common language is important because "partners need common representations, common interpretation and evaluation criteria in order to act together and interact in work situations" [DAV 09]. We use the library of generic models developed by [DAA 14] which allows modeling organizational scenarios and an ex ante performance evaluation (thanks to the simulation capability of the platform).

For this, we define the performance criteria. Interviews and discussion with representatives of the public sector, industrial sector and research institutions, enabled us to identify requirements on performance evaluation that are necessary to define them (Table 1).

Areas	Criteria
Socio-economic impacts	Production conditions
	SME's access to public contracts
	Environmental sustainability
Human resources aspects	Total number of jobs
	Total number of jobs lost
	Geographical distribution of jobs
Operational management	Global cost control
	Service quality
Change management	Process optimization
	Trend in demand
Audit & control	Maintenance of public expertise
Legal framework	Compliance with the financial clauses
	Compliance with service commitments
	Monitoring of litigation

Table 8.1. *Performance criteria for the "public clothing" function*

8.6. Conclusion

In this chapter, we propose a framework for extended administration. Our approach is based on identifying the specific interoperability requirements, according to six areas of management, to ensure organizational homogeneity for public–private networks. This framework allows organizations to capture the specific requirements of their network by adapting them to suit their own requirements regarding their direct partners in the network. Finally, it helps to clarify the requirements in terms of performance evaluation. We have experimented with this framework on this issue by defining the performance criteria for the "public clothing" function (Table 8.1). Future work would be to use this repository to evaluate reorganization scenarios for the public–private network of this function. It would also be to study the generic nature of this repository. Indeed, the criteria that make up the repository are not specific to the "public clothing" function.

8.7. Acknowledgments

The authors acknowledge CGI Business Consulting and the French National Association of Research and Technology (ANRT) which supports and partially funds this project. It is part of a partnership between the company CGI Business Consulting and the laboratory IRCCyN. The aim of this partnership is to provide methods and tools for extended administration.

8.8. Bibliography

[ATH 07] ATHENA INTEGRATED PROJECT, Guidelines and best practices for applying the ATHENA interoperability framework to support SME participation in digital ecosystems, Deliverable A8.2, 2007.

[AUB 02] AUBERT B., DUSSART A., Système d'information inter-organisationnel, Rapport Bourgogne, CIRANO, 2002.

[AUB 04] AUBERT B., PATRY M., Les partenariats public-privé: une option à découvrir, Rapport Bourgogne, CIRANO, 2004.

[AUB 05] AUBERT B., AZAMI L., BOURDEAU S. *et al.*, Synthèse critique d'expériences de partenariats public-privé, CIRANO, 2005.

[BEL 09] BELLAIS R., OUDOT J.-M., "Le recours aux contrats de partenariat dans la défense en France", *Revue française d'administration publique*, vol. 2, no. 130, pp. 263–274, 2009.

[BEZ 09] BEZES P., *Réinventer l'Etat. Les réformes de l'administration française (1962–2008)*, P.U.F., Paris, 2009.

[BLA 07] BLANC S., DUCQ Y., VALLESPIR B., "Evolution management towards interoperable supply chains using performance measurement", *Computers in Industry* , vol. 58, no. 7, pp. 720–732, 2007.

[BOU 90] BOURQUI M., "Relations interorganisationnelles et relations interpersonnelles: l'impact des réseaux sur l'approche concurrentielle des projets", *Revue Française du Marketing*, no. 127, pp. 51–61, 1990.

[CCI 11] CHAMBRES DE COMMERCE ET D'INDUSTRIE, Livre vert sur la modernisation de la politique de l'Union européenne en matière de marchés publics, available at: http://eur-lex.europa.eu/LexUriServ/LexUriServ.do?uri=COM:2011:0015:FIN:FR:PDF, p. 59, 2011.

[CHE 08] CHEN D., DOUMEINGTS G., VERNADAT F., "Architectures for enterprise integration and interoperability: past, present and future", *Computers in Industry*, vol. 59, no. 7, pp. 647–659, 2008.

[CHE 09] CHEN D., "Framework for enterprise interoperability", *8ème Congrès international de génie industriel – CIGI'09*, Bagnères de Bigorre, June 2009.

[DAA 11] DAABOUL J., Modélisation et simulation de réseau de valeur pour l'aide à la décision stratégique du passage de la production de masse à la customisation de masse, PhD Thesis, École centrale de Nantes, 2011.

[DAA 14] DAABOUL J., CASTAGNA P., DA CUNHA C. *et al.*, "Value network modelling and simulation for strategic analysis: a discrete event simulation approach", *International Journal of Production Research*, vol. 52, no. 17, pp. 5002–5020, 2014.

[DAC 07] DACLIN N., Contribution au développement d'une méthodologie pour l'interopérabilité des entreprises, PhD Thesis, University of Bordeaux 1, 2007.

[DAV 09] DAVOINE É., DEITMER L., "L'évalution participative et responsabilisante: un instrument de structuration des projets de partenaria publi-privé en recherche et développement", *Revue française d'administration publique*, vol. 2, no. 130, pp. 275–289, 2009.

[DUB 09] DUBÉ B., FACAL J., MAZOUZ B., "La communication, un outil de gouvernance des partenariats public-privé. Quelques enseignements tirés du cas Bonjourquébec. Com", *Revue française d'administration publique*, vol. 2, no. 130, pp. 291–306, 2009.

[EIF 03] EIF, European Interoperability Framework for Pan-European, European Communities, European Commission, 2003.

[HAF 09] HAFSI T., "Partenariats public-privé et management de la complexité: les nouveaux défis de l'Etat", *Revue française d'administration publique*, vol. 2, no. 130, pp. 337–348, 2009.

[HOO 91] HOOD C., "A public management for all seasons?", *Public Administration*, vol. 69, no. 1, pp. 3–19, 1991.

[IEE 90] IEEE, Standard Computer Dictionnary: A Compilation of IEEE Standard Computer Glossaries, IEEE, New York, 1990.

[INT 07] INTEROP, Enterprise interoperability framework and knowledge corpus, Deliverable DI.3, 2007.

[ISO 93] ISO/IEC 23812, Information technology vocabulary – Part 1: fundamental terms, ICS: 35.020; 01.040.35, 1993.

[ISO 98] ISO 14258, Industrial automation systems – concepts and rules for enterprise models, ICS: 25.040.01, 1998.

[ISO 00] ISO 17933, GEDI – Generic Electronic Document Interchange, ICS: 35.240.30, 2000.

[ISO 09] ISO 16100-1, Systèmes d'automatisation industrielle et intégration - Profil d'aptitude du logiciel de fabrication pour interopérabilité – Partie 1: Cadre, ICS: 25.040.01, 2009.

[ISO 11] ISO 11354-1, Advanced automation technologies and their applications – Requirements for establishing manufacturing enterprise process interoperability – Part 1: framework for enterprise interoperability, ICS: 25.040.01, 2011.

[LAB 04] LABROUSSE M., Proposition d'un modèle conceptuel unifié pour la gestion dynamique des connaissances d'entreprise, PhD Thesis, École centrale de Nantes, 2004.

[LEV 12] LEVESQUE B., "Social innovation in governance and public management systems: towards a new paradigm", in MOULAERT F., MACCALLUM D., MEHHMOOD A. *et al.*, (eds.), *Social Innovation: Collective Action, Community Learning and Transdisciplinary Research*, Edward Elgar Publishers, 2012.

[LOR 99] LORINO P., "A la recherche de la valeur perdue: construire les processus créateurs de valeur dans le secteur public", *Politiques et management public*, vol. 17, no. 17–2, pp. 21–34, 1999.

[MAL 11] MALLEK S., Contribution au développement de l'interopérabilité en entreprise: vers une approche anticipative de détection de problèmes d'interopérabilité dans des processus collaboratifs, PhD Thesis, University of Montpellier II, 2011.

[MAL 12] MALLEK S., DACLIN N., CHAPURLAT V., "The application of interoperability requirement specification and verification to collaborative processes in industry", *Computers in Industry*, vol. 63, no. 7, pp. 643–658, 2012.

[MAU 07] MAUCHAND M., Modélisation pour la simulation de chaînes de production de valeur en entreprise industrielle comme outil d'aide à la décision en phase de conception/industrialisation, PhD Thesis, École centrale de Nantes, 2007.

[MAZ 08] MAZOUZ B., FACAL J., VIOLA J.-M., "Public-private partnership: elements for a project-based management typology", *Project Management Journal*, vol. 39, no. 2, pp. 98–110, 2008.

[MAZ 09] MAZOUZ B., "Les aspects pratiques des partenariats public-privé. De la rhétorique néolibérale... aux enjeux, défis et risques de gestion des PPP", *Revue française d'administration publique*, vol. 2, no. 130, pp. 215–232, 2009.

[MED 13] MEDINI K., Evaluation de la performance d'entreprises dans le contexte de la personnalisation de masse durable, PhD Thesis, École centrale de Nantes, 2013.

[MEN 09] MÉNARD C., OUDOT J.-M., "L'évaluation préalable dans les contrats de partenariat", *Revue française d'administration publique*, vol. 2, no. 130, pp. 349–364, 2009.

[MOA 12] MOALLA N., PANETTO H., BOUCHER X., "Intéropérabilité et partage de connaissances", *Ingénierie des Systèmes d'Information*, vol. 4, no.17, pp. 7–17, 2012.

[OBS 12] OBSERVATOIRE ECONOMIQUE DE L'ACHAT PUBLIC, Gestion de la fonction habillement dans le secteur public, Ministère de l'économie et des finances, France, 2012.

[OSB 93] OSBORNE D., GAEBLER T., *Reinventing Government: How the Entrepreneurial Spirit is Transforming the Public Sector*, Penguin, New York, 1993.

[PAN 06] PANETTO H., "Rationalité du développement de l'interopérabilité", *8ème Congrès international de génie industriel – CIGI'09*, Bagnères de Bigorre, June 2009.

[PER 08] PEROTTI-REILLE C., Le textile habillement: la volonté de remporter la nouvelle révolution industrielle, Ministère de l'économie, des finances et de l'emploi, France, 2008.

[PIN 09] PINGAUD H., "Rationalité du développement de l'interopérabilité", *8ème Congrès international de génie industriel – CIGI'09*, Tarbe, 2009.

[RAU 14] RAUFFET P., DA CUNHA C., BERNARD A., "A dynamic methodology and associated tools to assess organizational capabilities", *Computers in Industry*, vol. 65, no. 1, pp. 158–174, 2014.

[ROC 10] ROCHET C., *Politiques publiques: De la stratégie aux résultats*, De Boeck, Paris, 2010.

[THO 86] THORELLI H.B., "Networks: between markets and hierarchies", *Strategic Management Journal*, vol. 7, no. 1, pp. 37–51, 1986.

[VER 07] VERNADAT F., "Interoperable enterprise systems: principles, concepts, and methods", *Annual Reviews in Control*, vol. 31, no. 1, pp. 137–145, 2007.

[WEB 63] WEBER M., *Le savant et le politique*, Union Générale d'Éditions, Paris, 1963.

9

An Inventory of Interoperability in Healthcare Ecosystems: Characterization and Challenges

A general design theory for interoperable socio-technical systems, as well as a control theory for such systems in interoperation, are still in progress. The diversity of interacting entities, their culture of collaborative practices, as well as the multiplicity of contexts and cooperation modes that they will meet during their life cycle, are all characteristics which hinder the rapid expansion of knowledge. Currently, many reference architectures contribute various solutions to make progress regarding this problem. They help to find the potential factors of complexity when facing interoperability requirements. At the very beginning of a project life cycle, they should help to address the right problem formulation in an efficient way and to speed-up the implementation of qualified engineering processes. Promoting interoperability in the specific area of eHealth is highly expected to enhance the quality and safety of patient care. It is particularly sensitive if an international domain is considered as a field of practice. Although eHealth interoperability system engineering is still considered to be in its infancy, several frameworks have already been published that are specifically designed for healthcare purposes. These frameworks help the designers to focus quickly and efficiently on the critical and/or relevant facets needed for the considered applications. Our objective is to introduce these initiatives and their motivations as a means to understand the specific features of the domain. Then, a survey and analysis of the main related work is developed. This analysis allows us to share concepts and discuss the major challenges in the near future.

9.1. Introduction

The use of eHealth solutions has a history of more than 30 years. It was first promoted for administrative and financial needs, but it has recently become a tremendous subject of innovation [CAT 11, RIA 07].

Chapter written by Elyes LAMINE, Wided GUÉDRIA, Ariadna RIUS SOLER, Jordi AYZA GRAELLS, Franck FONTANILI, Léonard JANER-GARCÍA and Hervé PINGAUD.

A comprehensive definition covering all facets of eHealth is found below: "*e-health is an emerging field at the intersection of medical informatics, public health and business, referring to health services and information delivered or enhanced through the Internet and related technologies(...)*" [EYS 01].

Indeed, as rightly pointed out by Pagliari *et al.* [PAG 05] "*In a broader sense, the term e-health characterizes not only a technical development, but also a new way of working, an attitude, and a commitment for networked, global thinking, to improve healthcare locally, regionally, and worldwide by using information and communication technology. So, Health and eHealth are inseparable due to the fact that health is an information intensive sector*".

More specifically, as depicted in Figure 9.1, eHealth promotes access to information for a large panel of professionals and patients and facilitates cooperation between actors of an ecosystem (see circles in the middle of the figure). It also helps to develop new modes of healthcare (see the business cases in the upper part of Figure 9.1), in order to increase the number of services for patient care [BLA 11]. Many new applications and information and communication technologies have been specifically designed or simply adapted for eHealth (see the lower part of Figure 9.1).

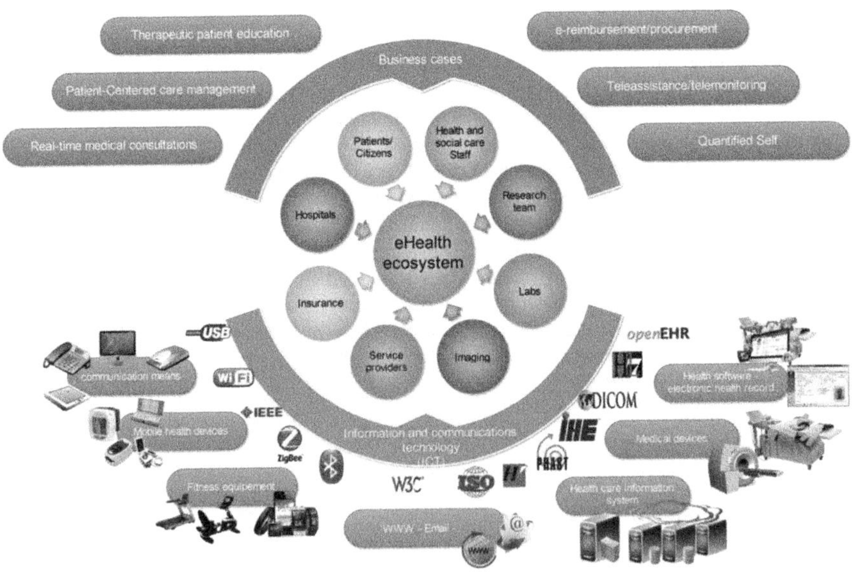

Figure 9.1. *eHealth ecosystem*

Logically, the emergence of new information and communication technologies, such as the ones related to adaptations of Web 2.0 innovations, fosters the development of eHealth. However, related innovations have not only been technological, the application of Web 2.0 in healthcare has led to the emergence of a new paradigm, Health 2.0, introduced in the mid-2000s, which characterizes a transition of the patient status from passive consumer profile to an active participant in his/her health. This new motivation has had strong effects on public healthcare policies. For example, it increases the need for chronic illness data in the ecosystem. Simultaneously, the move toward continuous patient health consideration has been correlated with a change in individual attitude, making prevention an increasing reality. Rather than belated awareness linked to the diagnosis, proactive wellness gradually became a preference. All these effects have definitely influenced the way healthcare should be managed as a whole.

With a little hindsight on the phenomenon, two types of benefits are considered to be related to the use of eHealth services and tools. In a general perspective, they can be summarized as:

– soft benefits for patients: related to wellness, improved relations between patients and healthcare professionals, increased patient comfort;

– hard benefits for the healthcare management system: dealing with cost reduction and easier access to care; some of the performance indices being the reduced number of hospital admissions and trips between home and hospital for the patient, the smaller number of emergency hospital admissions and decreased levels of hospital re-admission.

Even if these benefits have been widely recognized, we observe that the wide-scale adoption of eHealth is still slow. According to the latest 2014 eHealth Initiative survey on health data exchange [EHE 14], one of the main limiting factors is the lack of interoperability between operational systems, which is considered as the most significant cause of eHealth project failures.

We actually know that eHealth solutions are mainly designed with key success factors in mind, such as openness and sharing properties. The reasons behind this lie in the fact that technologies meet a very limited range of needs within a medical operating protocol. A solution has to be frequently included within a collection of support resources of other protocols. And here is the main problem. The ecosystem is *de facto* populated by new technological inputs that have not been systematically designed in the perspective of mutual interactions. Each solution offers a service delivered to its end user considered as a target market, fostering time to market constraints. However, in practice, integration shall be effective, with minimal effort, during the protocol execution. This is why interoperability becomes attractive, as it provides the capacity to interact seamlessly to many entities.

As the situation has been noted from the field study, it seems necessary to study the research advances in the domain and the purpose of existing initiatives on eHealth systems' interoperability, and then our ability to use this knowledge to detect the current shortcomings and address them. These are the main motivations of this chapter.

The remainder of this chapter is organized as follows. Section 9.2 deals with the eHealth interoperability research context and its benefits and challenges. Section 9.3 gives an overview of the different meanings of the concept of interoperability in practice. In section 9.4, we compare different interoperability frameworks for the health domain and highlight their relevance as well as their coverage with regard to the eHealth interoperability domain. Finally, in section 9.6, we conclude and present some future works.

9.2. eHealth interoperability

One of the oldest definitions of interoperability is the one proposed by Jay and Goetz [JAY 88] considering interoperability as the "*ability of two or more systems or components to exchange information and to use the information that has been exchanged*". Here, the focus is put on fluent communication between systems, mainly in a technical perspective. One of the main drawbacks of the IEEE definition is that it does not help to make the difference between the concepts of system integration and system interoperation.

A more evolved definition proposed in Pingaud [PIN 09] prefers apprehending interoperability as an organizational challenge by referring to system engineering. For the author, interoperability is the "*capacity of natively independent systems to interact properly, i.e. to behave in harmony, sharing easily a common objective, without having to deeply modify their own structure and behaviour*". The point of view chosen here is neither limited to the nature of the exchange nor restricted to ICT, but concerns a common mission that the contributors have accepted to perform and the energy they will use to execute it. The term harmony is used to describe the data, the software services, the processes, as well as the business activities and related practices that will be involved during a collaboration. As a consequence, the definition extends the interoperability concept to a rational approach of the organization.

From the same perspective, the European Commission[1] has defined interoperability as the "*the ability of disparate and diverse organisations to interact towards mutually beneficial and agreed common goals, involving the sharing of*

1 Decision No 922/2009/EC of the European Parliament and of the Council of 16 September 2009 on interoperability solutions for European public administrations (ISA): http://eur-lex.europa.eu/legal-content/EN/TXT/?uri=CELEX:32009D0922

information and knowledge between the organisations, through the business processes they support, by means of the exchange of data between their respective information and communication technology (ICT) systems".

As outlined in their official publication [IDA 08], interoperability is often confused with other closely related concepts, such as:

– Integration – the act of forming a functional system from loosely coupled components able to work together as a whole, components for which it must be pointed out that they have been natively engineered with this objective in mind;

– Compatibility – the interchangeability of tools in a particular context, which is often assimilated as a passive property during design, making things possible, but not necessarily performing them;

– Adaptability/flexibility – the ability of a system to adapt its structure or behavior to changes in its environment, which is basically a dynamic property that is difficult to understand and is exploited in the context of interaction between many contributing systems. Adaptability/flexibility are system properties that could be reached easily by making use of interoperability.

In the healthcare context, the National Alliance for Health Information Technology defines interoperability as "*the ability of different information technology systems and software applications to communicate, to exchange data accurately, effectively, and consistently, and to use the information that has been exchanged*" [HEU 06]. The definition deals somewhat with a cause-and-effect relationship: the data exchanged by the communication facilities are useful and, as so, should be interpreted by the receiving party. It is a drastic simplification of the concept that minimizes the goals and organizational features, prioritizing data exchange. Once again, if such a situation is desired, the fact is that it could be reached by an integration approach (which is not equivalent to interoperation).

The CEN/ISSS eHealth Standardization Focus Group defines interoperability along the same lines as "*a state which exists between two application entities when, with regard to a specific task, one application entity can accept data from the other and perform that task in an appropriate and satisfactory manner without the need for extra operator intervention*" [CEN 05a].

In general terms, eHealth interoperability, as defined by the institutions cited above, means a technical contribution mainly operating at the software and data layers of the global system architecture that could provide benefits to actors during some of their collaboration acts. This understanding clearly tends to forget many of the specific constraints that are real barriers to eHealth interoperability at least at the business layer of the architecture, even far beyond the architecture, at the scale of health policies and systems' governance rules. Generally recognized as a good practice in IT management, alignment of the IT system onto the corporate system is a prerequisite to a high service quality. Thus, following roughly the same consideration, we have to go back to a characterization of the eHealth domain, considered as a set of collaboration spaces, to overcome this limit.

A preliminary analysis of contexts in which interoperability is needed will enrich the approach of interoperability concerns. Indeed, the continuous improvement of patient healthcare, achieved by improved awareness, accessibility, coordination and continuity of care, requires better management of information flows with real added value for the system. Similarly, cooperation between stakeholders of the healthcare management processes is more and more intensive, leading to an increasing n umber of collaboration opportunities between them. This was the basis for the consideration of a healthcare system as a human centric supply chain or value chain, where actors around the patient are assimilated to healthcare service providers. The CEN/ISSS eHealth Standardization Focus Group has identified such strategic aspects when it specifies: "*interoperability is the only sustainable way to help partners acting in various locations, with different expertise, perspectives, statuses and agendas, possibly cultures and languages, and using distinct information systems from different vendors, to collaborate harmoniously to deliver quality health care*" [CEN 05b]. Although these interoperability requirements are recognized, among others, as a cornerstone for improving the quality of healthcare and the efficiency of the overall health system organization, it is difficult to explain them simply without considering the multidimensional aspects of the concept. Some of these aspects are attached to value chain features that have to be considered and will go beyond the technical analysis.

The healthcare system fulfills its missions within the perimeter of a known territorial scope and mainly in the form of closeness to citizens. Indeed, it is within the field of public policy. National or international cells (political community) prevail. The society is logically expecting the best of healthcare services. Public policies elicit the design of the operational part of the system as well as setting up its managing rules. The design is based on multiple aspects relative to many independent components that are involved in healthcare services. Starting from a description of each aspect, we can draw the perspective of potential interactions between these components:

– Organizational: adapt public health reglementation and recommendation at a regional level, classification of entities authorized to supply care and the care services including drug administration, patient community of interest or community of interest to practitioners. We are at the core of governance with a rationalization of healthcare services functionally defined and performed by competent institutions able to provide them;

– Cultural: social constructs having a historical basis, showing the habits and practices of patients, laying their consumer demands for health services and fixing the accepted tolerance levels on the quality of these services, for addressing the form and nature of relations established between practitioners and patients, particularly around their modes of communication, but also between patients or between practitioners either individually or collectively. Culture also refers to knowledge and skills and how to disseminate knowledge, including training of health professionals. Finally, culture also covers the ways of life of a population and, therefore, the relationship between living conditions and health status of a population;

– Ethical: a sum of rules sets the range of acceptable behavior by healthcare practitioners; they define them following a corporatist compromise. These rules are related to their practices, what to do and what not to do, what to say and what not to say. Therefore, they set the framework within which the exchanges between actors should happen. The ethics help in the recognition of the status of a profession in a society. The advent of new health devices, whether technological or not, raises ethical issues because the rules must change;

– Economical: the health system is a service that has a cost. The management of this system obeys the rules based either on a liberal and direct relationship between patients and practitioners or on a link by a tier, health insurance, which provides coverage of expenses of the insured patient. Depending on the country, health insurance can be a private entity, a public one or a mixed structure. Nevertheless, access to care is achieved by an economic model that does not escape the law of supply and demand. Care services use the appropriate resources available in a nearby environment. Accessibility is a criterion for judging the performance of a system, in which the patient should be treated at the right time, at the right price, in the right place and with the right quality;

– Financial: the structure of a healthcare system is based on tangible and intangible investments. These investments consume financial resources. The availability of resources and their qualifications are the foundation of equity for citizens. Here again, the origin of the investment funds may be public or private. The return on investment can be judged against several criteria describing the service, criteria which are both economic and social. Because of the size of a health system, as well as its strategic importance, the system raises sustainability issues. The relevance of the investments is a highly sensitive point, referring to a strategy that has to be defined and widely accepted. It affects the economic equilibrium of the system during operation;

– Legal: public health policy is defined by the legislation. These legal elements include the preservation of individual rights, establishment of the competencies and duties of official entities, providing a legal framework for the exercise of practitioners, the use of which leads to circumscribe the scope of responsibility and finally often frames the economic model to an institutional level so that citizens' expectations are met.

Each of these aspects is an inducer of information or knowledge on the structure and behavior of the healthcare system. Information flows can not only be medical or social, but also financial or legal. Each of these aspects is a crucible that can lead to the birth of collaborations. The definition of a collaborative context can rarely ignore these different aspects that are taken into account in the analysis of requirements for interoperability between system components. The dynamic behavior of this system is the result of changes in each of these areas over time. The health system is a living system that is constantly evolving to improve its performance and learning from its own experiences. Of course, different time constants characterize this dynamic. Consequently, the need for interoperability has no basis outside a certain time pressure and/or a need for organization flexibility to face an unpredictable context.

Hence, interoperability concerns have become a priority for eHealth within many countries, as a visible part of their political strategies. Some of them have developed entire institutions into which interoperability progress is defined as a prior mission. In the following, we provide some examples: Australia's *National E-Health Transition Authority* (NEHTA) was established to lead the progression of eHealth in Australia, with the goal of creating a sustainable and interoperable eHealth environment. The same situation is a reality in France, where *ASIP Santé* is the official public agency in charge of eHealth development. Practically, ASIP defines and promote frames of reference, standards, products and services, which contribute to the interoperability, security and usage of health information systems and tele-health. In the same way, the *Healthcare Information Technology Standards Panel* in the United States is in charge of developing standards for local, regional and national health information networks. In Catalonia, the *Oficina d'Estàndards i Interoperabilitat de la Fundació Tic Salut* is the National Agency for these purposes.

9.3. Levels of interoperability in eHealth ecosystems

In the following, we adopt the EHR Interoperability Work Group view as the main starting point for considering interoperability as a noun describing mainly fluent relationships between independent systems rather than a quality or a qualification [GIB 07]. The related relationships can be specified and categorized at different levels in a healthcare context, such as the interoperability of the exchanged messages between healthcare applications (information level), interoperability of Electronic Healthcare Records (services level), interoperability of patient identifiers (functional level), interoperability of healthcare business processes (process level) and interoperability of codified terms (technical level). However, all these categories of interoperability can be merged into three main levels: technical interoperability, semantic interoperability and organizational interoperability [BRA 12, COC 15, GIB 07].

9.3.1. *Technical interoperability*

Technical interoperability is usually focused on the technical issues related to the connectivity across the network and applications, and also to specific context-aware communications by well-structured messages. Historically, technical interoperability inherits knowledge coming from the area of electronic data exchange. It is often centered on the definition of standards and protocols needed to insure the exchange and sharing of data among the seven layers of the OSI reference model [KUB 11]. However, it is important not to neglect the specificity of eHealth interoperability, which is essentially concentrated on message coding and decoding [GIB 07].

In the healthcare domain, many approaches deal with the data structure of the messages that will be exchanged between known parts. Several standards and norms of exchange and data sharing have been developed to enable systems and applications often designed "in silo" to be interoperable by sharing a common data structure. For a linguist, this normative literature will provide the prescription of a syntax in a common language shared by the actors during communication.

Among these norms and standards, we here describe:

– HL7 (Health Level Seven), a standard dedicated to provide a comprehensive framework for the exchange, integration, sharing and retrieval of electronic health information that supports clinical practices and the management, delivery and evaluation of health services [BEN 12];

– DICOM (Digital Imaging and Communication in Medicine), an open standard managed by an international committee (the DICOM Committee), for handling, storing, printing and transmitting information in the area of medical imaging [PIA 09];

– PN13-SIPH, a standard resulting from the work of Interop/SIPH group within the Phast association. This standard focuses on the exchange of information directly related to drug supply: from the prescription up to the drug supply and use [BOU 14];

– H.PR.I.M (*Harmoniser et PRomouvoir l'Informatique Médicale*), a standard for biology exams transmission between laboratories, health organisms and prescribers [ROT 93].

In addition to these multiple norms and standards, which widely cover the health domain, there are various instances that encourage stakeholders to share and make use of interoperability standards. Among these instances, we here describe:

– IHE[2] (Integrating the Healthcare Enterprise), an international initiative by healthcare professionals and industry to improve the way computer systems in healthcare share information. IHE promotes the coordinated use of established standards, such as DICOM and HL7, to address specific clinical needs in order to support optimal patient care. Systems developed in accordance with IHE communicate together in a better way, are easier to implement and enable care providers to use information more effectively;

– Continua Health Alliance[3], an international not-for-profit industry organization enabling end-to-end, plug-and-play connectivity of devices and services for personal health management and healthcare delivery. Its mission is to empower information-driven health management and facilitate the incorporation of health and wellness into the day-to-day lives of consumers.

These norms and standards, as well as the communities that support them, have reached a high level of maturity. Indeed, Figure 9.2 illustrates a very common configuration in a health information system architecture, in which we can clearly see that exchanges between the various components of these information systems are based on the use of multiple known standards and designed especially for the health domain. For example, we will use the DICOM standard to store radio images in the PACS and HL7 standard to feed the patient record, stored in the hospital information system, with data from different servers, particularly from PACS and medical laboratory servers.

However, the contrast between the multiple standards contributing to interoperability and the lack of empowerment of actors in the health field is still clear. Thus, for instance, exchange formats often belong to one provider, for whom it is easy to apply a "customer retention" strategy, building systems that are only interoperable within the perimeter of their own products for captive customers. Moreover, the achievement of technical interoperability in healthcare is pragmatically challenged by the clarity and consistency of recommendations delivered by authors in a given release of a standard (which requirement is mandatory and which is not). Considerable effort is required in terms of energy and time in order to choose the adapted standards, as well as to implement a standard compliant solution in a given context. Any misunderstanding in the recommendations could affect the compliance level of a standard by a given actor and lead to severe consequences for the overall system as a butterfly effect.

2 http://www.ihe.net/

3 http://www.continuaalliance.org/about-continua

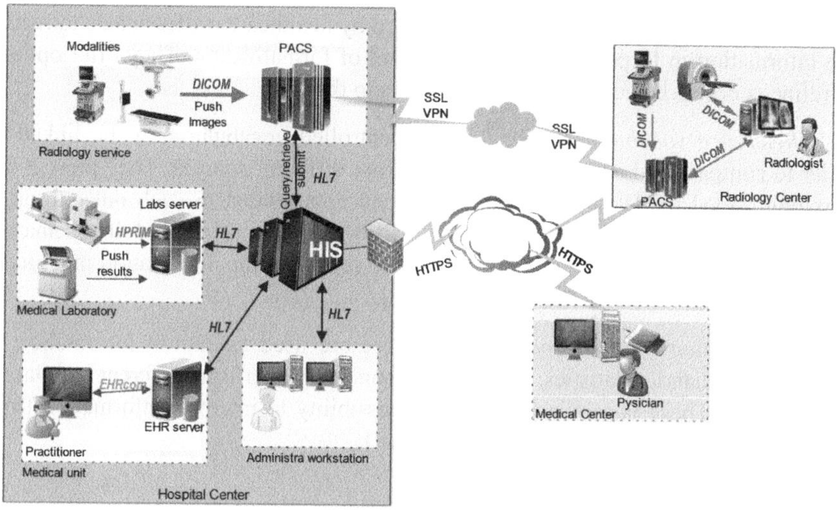

Figure 9.2. *Use of well-known standards specifically designed for health sector*

9.3.2. *Semantic interoperability*

Semantic interoperability focuses on how to manage information flows coming from or going outside a collaborative entity as a source of knowledge.

Semantic interoperability is the ability to exchange information between components, allowing its use in the receiver as if it had been generated there. This level of interoperability ensures the correct interpretation of the exchanged content without ambiguities, even when multiple languages are used. The objective of semantic standards is to represent the meaning of the recorded information by assigning it an unambiguous code that can be automatically processed in Information Systems (ISs). The encoding of the information using one of these standards implies the normalization of its meaning by establishing the common descriptions, but also its representation using the specific conceptual model in which the standard is based. Once clinical data have been structured, we can search, query, explore, compare and exploit them.

In the healthcare domain, *natural language* is the oral or written language used by healthcare professionals to communicate with each other and with their patients. It is very flexible and specialized, but it can be ambiguous too as it is fully context-dependent. Natural language cannot be used by ISs to process information automatically, but it is necessary to structure it. *Controlled vocabularies* are specifically designed to represent information in IS, and we can classify them into three different types, as we can see in Figure 9.2:

– *Interface (input) vocabularies*: these used by healthcare professionals to record the information in IS performing the first level of formalization. They offer options to refine concepts by adding more information to them;

– *Reference vocabularies*: these allow controlled vocabularies to be linked in order to compare and consider information from different sources. They present the maximum level of detail and should have concepts related to each other through different types of relationships. They allow us to achieve fully semantic interoperability of contents, and can be used to represent the information in Clinical Decision Support Systems (CDSS) [AHM 11];

– *Aggregation (output) vocabularies*: these are used to exploit information in order to plan, handle statistics, manage reimbursement, perform different population analysis, etc. These vocabularies offer the possibility to interpret information from different viewpoints (e.g. clinical and financial).

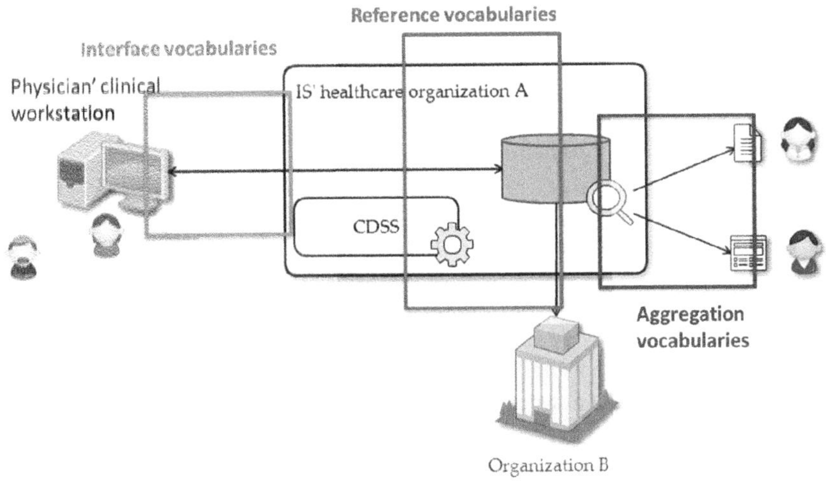

Figure 9.3. *Types of controlled vocabularies*

Focusing on the objective and the structure of controlled vocabularies, they can also be classified into the following types:

– *Thesaurus*: they are formalized lists of terms (each of them with associated keywords) related to each other through hierarchies, equivalencies or associative relationships (like synonyms of the same idea). They are designed to index or mark content in order to facilitate its recovery in documents, articles or other databases;

– *Catalogues*: they are plain lists of codes with descriptions that can have more attributes. Their objective is to facilitate the identification and localization of information using pre-sorted elements;

– *Classifications*: these vocabularies are made of elements with codes and descriptions that can be grouped in chapters and that are organized in classes and subclasses. They follow a tree hierarchy structure modeled by codes that contain different levels of detail, including miscellaneous and unspecific classes. They aim to classify the information of patients;

– *Terminologies*: these vocabularies are comprised of terms that are normally related to each other through a certain hierarchy and other types of relationships. Their elements can have more than one description and many other attributes. They can have concepts from many domains. Their objective is to relate concepts from different sources and controlled vocabularies that represent the same idea using the maximum level of detail;

– *Ontologies*: they are semantic networks of concepts with a high level of abstraction and are related to each other using different types of associations. Their elements can have many attributes and can be encoded using a terminology. Their objective is to represent meanings in order to allow for their automatic processing.

Table 9.1 shows the correspondence between these two classifications of controlled vocabularies.

Type of resource	Type of vocabulary	Examples
Thesaurus	Aggregation	COSTART, MeSH and DeCS
Catalogues	Interface	UCUM, SERAM and the Spanish catalogue of countries
Classifications	Interface and aggregation	ICD-10, ATC and ICPC-2
Terminologies	Interface, reference and aggregation	SNOMED CT, LOINC and MedDRA
Ontologies	Reference	OpenGalen, GO and FMA

Table 9.1. *Correspondence between classifications of language*

One of the most complete and widely used reference terminologies is *SNOMED CT*. This international standard provides a homogeneous base of representation for different healthcare domains (it is organized into 19 axes that include clinical finding, procedure, substance and social context) that are connected through different types of relationships (e.g. "*has active ingredient*", "*has dose form*", "*IS_A*", "*pathological process*", "*due to*"). It allows for the use of synonyms and supports multilingual descriptions that define each concept in natural language

[IHT 14]. SNOMED CT can be used as an interface, reference and aggregation terminology because it can be used to: (1) record the information of patients from clinical workstations by healthcare professionals; (2) compare data from different sources and (3) analyze the encoded information. When we use SNOMED CT as reference terminology, we can still continue using other controlled vocabularies in our ISs and at the same time we can take advantage of the standardization in the exchange of information, mapping these (local) vocabularies to SNOMED CT.

Healthcare organizations use many types of controlled vocabularies with different structures and objectives, and therefore they need to create, represent, update, distribute, explore, translate and relate them. Managing these semantic ecosystems is one of the main challenges in the semantic interoperability sector, and good tools specially designed to cover all these functionalities are needed. *Terminology servers* are the type of tools that cover these requirements as they allow organizations to centralize the management of controlled vocabularies, its components and mechanisms, decoupling the process from other ISs. The interaction between terminology managers and the terminology server is done through a (web) user interface.

The difficulty in using these semantic resources to encode the information recorded by healthcare professionals is the other main barrier. We need to ensure that these data are stored in a structured way but without it being necessary for physicians to work directly with codes. To achieve this objective, we can combine usability techniques with terminology services that facilitate the recovery, search and query of controlled vocabularies from clinical workstations. Terminology servers provide this type of (web) service,[4] and we can use them from clinical workstations to recover descriptions in natural language that have associated standard codes (as they are part of a controlled vocabulary). With this approach, physicians can work with understandable descriptions, and at the same time normalized codes are stored. The use of terminology servers is currently growing and the number of providers of this type of solution is growing from day to day.

9.3.3. *Organizational interoperability*

Organizational interoperability focuses on how to manage information flows, or going outside a collaborative entity, as signals for achieving individual goals, delivering expected services, executing private or public business processes and partners' needs.

4 HL7 has developed a standard called CTS2 that provides a functional definition of the operations as services that can be applied to semantic resources and provided by a terminology server.

Organizational interoperability is the third part of the concept that is achieved in enterprise engineering "defining business goals, modelling business processes and bringing about the collaboration of administrators that wish to exchange information and may have different internal structures and processes" [IDA 08]. It is mainly related to the definition of business interfaces across several organizations and into a collaborative space. It also deals with the way of assigning responsibility and authority to perform a given process [CHE 07]. It has been identified as a requirement to actually achieve a specific delivery service among involved players and business processes. Indeed, in Vernadat [VER 10], organizational interoperability is considered as the ability of business organizations to provide services to each other as well as to users or customers, or beyond that, to a wider public in the case of administrative organizations.

In the context of the healthcare domain, organizational interoperability is still rather difficult to depict exhaustively even if a brief description of actors has been provided in section 9.1. In fact, our own experience on this subject as well as available knowledge in the current literature set out the use of three terminologies, namely process interoperability, business interoperability and organizational interoperability, which are similar to one to another. In our opinion, the term organizational interoperability is a broader concept that includes the two others and provides a better meaning for a large public.

According to Telemedicine Alliance (TMA) [I2H 05], the organizational interoperability is considered as the state where the organizational components of the health system are able to perform seamlessly together. The vision is an integrated health system that provides efficient, effective and holistic citizen-centered services on the basis of the principles of Health for All-Access, Equity and Solidarity.

In the healthcare domain, such an interoperability level deals essentially with individual services able to improve the collaborative work and help to exchange information between the different stakeholders involved in care process executions. For example, patients' identity matching, results delivery, summary of care document sharing and "recovering and payment orders" are proven helpful services and are stepping stones on the way to managing care pathways. However, although this service view is useful to fix the contribution of everyone to the collaboration, a process view shall be superposed if we want the work to be performed collectively. This new step is often described as a coordination act between actors. In this event oriented approach of organizational interoperability, the aim will be to provide users with abilities allowing them to perform their activities at the right time and to connect properly to other services in a time-, event- or sequence-oriented manner. It is worth noting that these requirements are also needed to be addressed in the new type of numerical relationships between citizens and healthcare professional organizations.

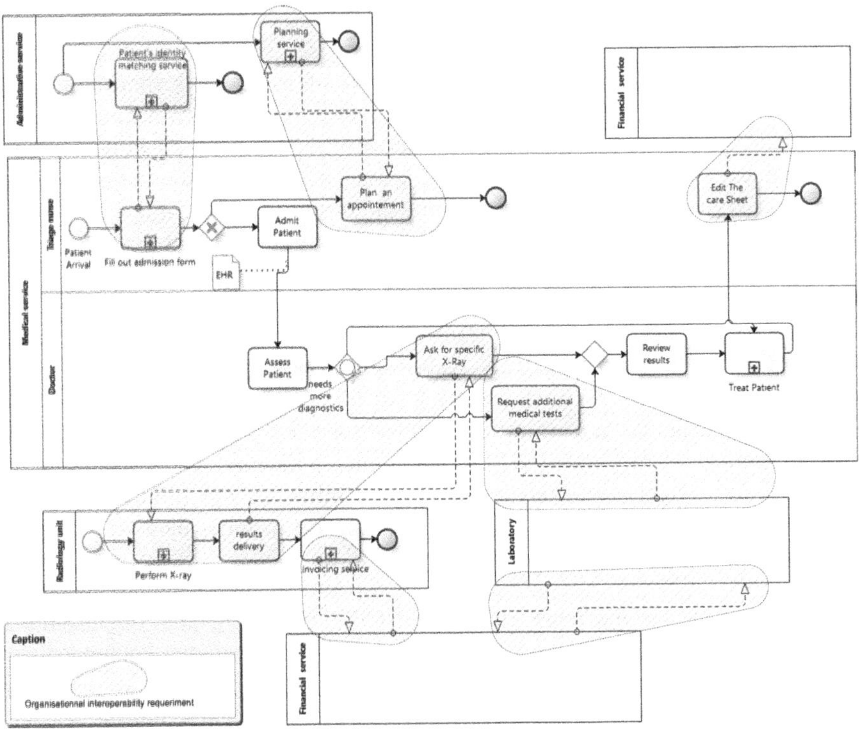

Figure 9.4. *Business process considering the diagnosis for a patient in a hospital-event-oriented view*

In Figure 9.4, the BPMN language is used to model the business process for a patient diagnosis in an outpatient care during a visit to the doctor at a hospital. The organization is described by the pools and the lanes, which identify the roles of the different actors. The BPMN language is mainly oriented toward the sequence of activity leading to the final result with respect to time. The dynamic of process execution fulfills the need for coordination of activities performed by relevant actors in order to reach the final goal. The triggered events point out the activities that could begin; the occurred event specifies the one that will be launched and the next set of events that shall be supervised. Obviously, the information flows between parts are inputs or outputs of activities, controlled in running phase by interactions between applications embedded in the execution of activities. This can be supervised by a workflow engine, for example. It will help to connect different types of interfaces like software interfaces and/or human–machine interfaces.

Starting from this level of analysis, one important practical feature in existing solutions is the "profiling approach". It consists of defining "building blocks", called profiles that publish information about the interfaces of an entity. Profiles are promoted by standards and norms where they are described as meta-structures containing all the needed knowledge for numerical interaction with others. They address a broad range of aspects, covering data, service and process specifications. It could consider many things: information transport, security, data structures in messages and related data models, associated terminologies, privacy and service contracts [RHO 10]. It should not rule on how these communicating systems should be designed internally but focus on their engagement as interoperating actors in the context of the profile. Profiles are useful aggregates in interoperability engineering, as they provide all the informational assets of a given system in a consistent way. Hence, the main objective of organizational interoperability will be to effectively assemble and combine the right set of profiles to address a real business use case.

To facilitate and promote the use of standards and the profiling approach by eHealth providers in Europe, the European Commission in 2007 addressed the mandate M/403 (also called eHealth Interop) to several standardization organizations (ESOs, CEN, CENELEC and ETSI). The challenge was to develop a coordinated program for improving interoperability in health informatics. The aim of the M-403 is to provide a consistent set of standards and best practices to address the needs of the eHealth field for the benefit of future healthcare provision [REY 09]. It does not consist of creating more standards for IT, but encouraging agreement on which existing standards should be adopted to ensure interoperability between health IT systems at an international level. Indeed, in this work program, eHealth-INTEROP holds the view that interoperability is not a destination in itself that can be set up and defined by this working group, but it is a road that needs to be taken properly by many different types of stakeholders across many jurisdiction spaces. To this end, the M/403 proposes a coordination process to complement the current standards development processes so as to meet the expected business use cases by delivering profiles specifications and the associated tests able to verify the compliance of the interoperability efficiency.

This process was mainly composed of five major activities as depicted in Figure 9.5:

1) use case definition and prioritization deals with the identification and prioritization of needs expressed in the proposed business use cases;

2) base standards development and maintenance consists of checking the adequacy of the existing standards panels regarding the needs expressed above;

3) profile development and maintenance, where the main task relies on the consistent assembly of selected standards to ensure the business use case prioritized setting;

4) quality assurance of profile implementations is responsible for the development of conformance test plans and test tools able to verify the quality of the adopted solution. This process is usually performed by IHE (for eHealth);

5) best practice forum, this aims to give every national or regional project a forum to share experiences and publish best-practice guides that offer guidance in deploying eHealth project activities.

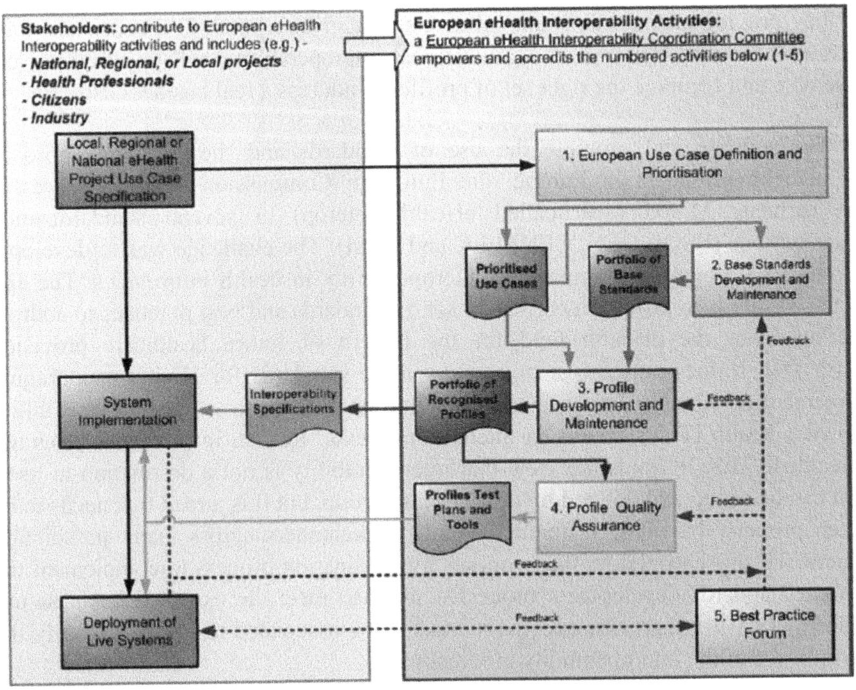

Figure 9.5. *Key activities to achieve eHealth interoperability [REY 09]*

9.4. Survey of interoperability frameworks

Developing an interoperable system or control of a system in interoperation should obviously be considered from multiple perspectives, various dimensions and with different types of approaches.

Several research studies and initiatives have focused on the structured approach of interoperability engineering. The main dimensions of interoperability have to be emphasized and have to be inserted in a consistent framework that provides best practices for projects and contribute to the progress of knowledge.

In this section, we introduce and discuss the main interoperability models and frameworks relevant for healthcare applications.

9.4.1. *eHealth European Interoperability Framework (eHeath EIF)*

The eHealth EIF framework [EUR 13] was developed in the context of a research program funded by the European Commission and focused on interoperability development. It is an application of the generic European Interoperability Framework (EIF) [EUR 11] to the domain of eHealth. eHealth EIF aims at providing a set of recommendations and specifications to connect eHealth systems. It identifies four levels of interoperability: legal, organizational, semantic and technical. An overview of these different EIF concepts is given in Figure 9.6:

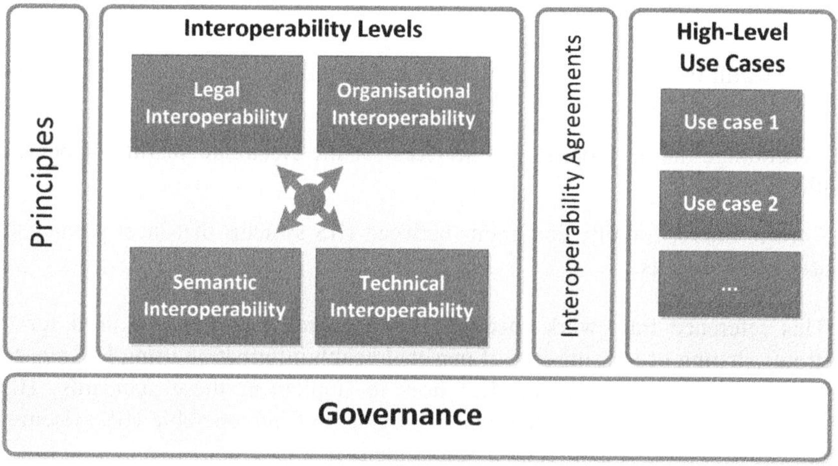

Figure 9.6. *Structure of the eHealth EIF [EUR 13]*

– legal interoperability aims to "align legislation so that exchanged data are accorded proper legal weight";

– organizational interoperability aims to "coordinate processes in which different organizations achieve a previously agreed and mutual beneficial goal";

– semantic interoperability aims to specify the "meaning of exchanged information, which is preserved and understood by all parties";

– technical interoperability aims to "discuss technical issues involved in linking systems and services".

For each one of these interoperability levels, the involved organizations must formalize cooperation agreements. Interoperability governance "covers the ownership, definition, development, maintenance, monitoring, promoting and implementing of interoperability frameworks in the context of multiple organizations working together to provide (public) services".

Six principles were defined on the basis of the generic EIF: security and privacy, transparency, preservation of information, reusability, technological neutrality and adaptability and openness. Two additional principles have been included in this list: patient centricity and an approach based on use cases. The eHealth EIF proposes a list of 10 high-level use cases to the eHealth Network.

9.4.2. *Health Information Systems Interoperability Framework (HIS-IF)*

The Health Information System (HIS) Interoperability Framework is a reference framework created by ASIP Santé[5] to:

– encourage the development of services for the electronic sharing of personal health information;

– create interoperability conditions between HIS systems that meet privacy and security requirements.

This reference framework specifies the standards that must be used for the electronic sharing or transmission of personal health information using HIS systems [ASI 10]. The model also specifies how to implement these standards (HL7, DICOM, etc.) in order to facilitate the deployment of interoperable HIS systems in agreement with privacy and security requirements.

The HIS Interoperability Framework (HIS-IF) reference is divided into modules, which are distributed across three interoperability layers:

– content layer (semantic and syntactic content): specification of exchanged or shared content in terms of structure and vocabularies;

5 *Agence nationale des Systèmes d'Information Partagés de Santé - http://esante.gouv.fr/en*

– service layer: specification of content sharing or exchange services, their rules and usage parameter;

– transport layer: specification of exchange protocols used by services.

The service and transport layers are often referred to as a group as the "Technical Base" of the framework, because both are developed in a technical development track with vendor participation. The content layer is mostly focused on user input and requirements and developed separately [ASI 10].

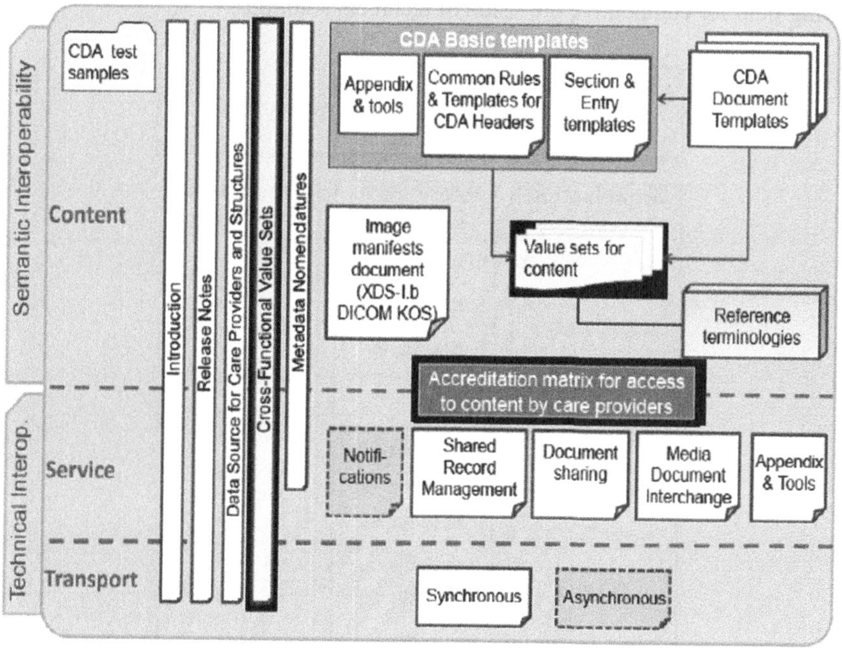

Figure 9.7. *Organization of the HIS interoperability framework [ASI 10]*

9.4.3. *eHealth Interoperability Framework (eHealth IF)*

The eHealth interoperability framework [NEH 07] was developed by the National E-Health Transition Authority (NEHTA[6]) initiatives in Australia. It defines three levels of interoperability across health organizations (Figure 9.4):

6 http://www.nehta.gov.au/about-nehta

– organizational layer, which provides a shared policy and process framework across the eHealth interoperability agenda covering each NEHTA initiative. It includes the business processes, standards plan, security policies and privacy issues;

– information layer, which provides shared building blocks for semantic (information) interchange, including foundation components, value domains, structures, common assemblies, relationships and metadata;

– technical layer is concerned with the connectivity of systems for information exchange and service use. Solutions are based on open standards providing a level playing field for competitive provision of technical solutions.

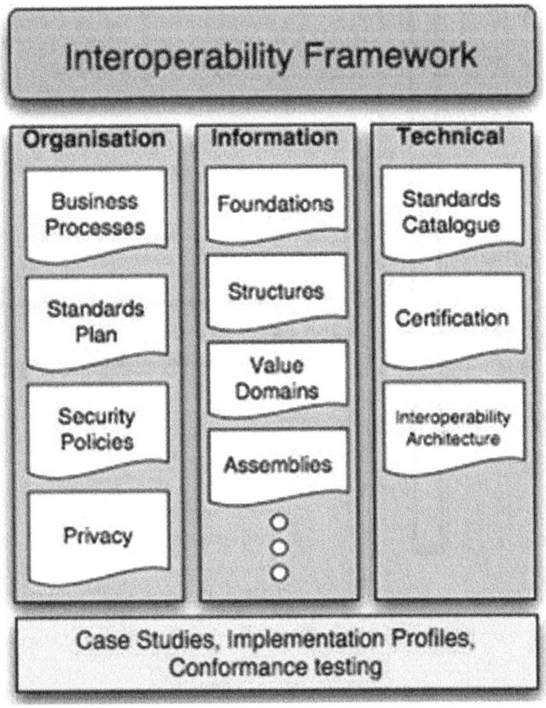

Figure 9.8. *eHealth interoperability framework [NEH 07]*

9.4.4. *Personal Health Systems framework*

Personal Health Systems (PHS) assist in the seamless provision of quality-controlled and personalized health services to individuals, regardless of their location. They consist of: (1) *ambient and/or body devices* (wearable, portable or implantable), (2) *intelligent processing of the acquired information and coupling of it with expert biomedical knowledge* to derive important new insights about an individual's health status and (3) *active feedback* based on such new insights [EUR 11].

As can be seen in Figure 9.6, the PHS Interoperability Framework (PHS IF) can be divided into two smaller frameworks: (1) *technical and implementation framework*, including standards, profiles and guidelines for their implementation based on elaborated business use cases, identification and authentication mechanisms, security protocols, testing and certification, etc. and (2) an *institutional/organizational framework* encompassing policy issues (e.g. governance, reimbursement) and legal and regulatory aspects such as data protection and liability [STR 11].

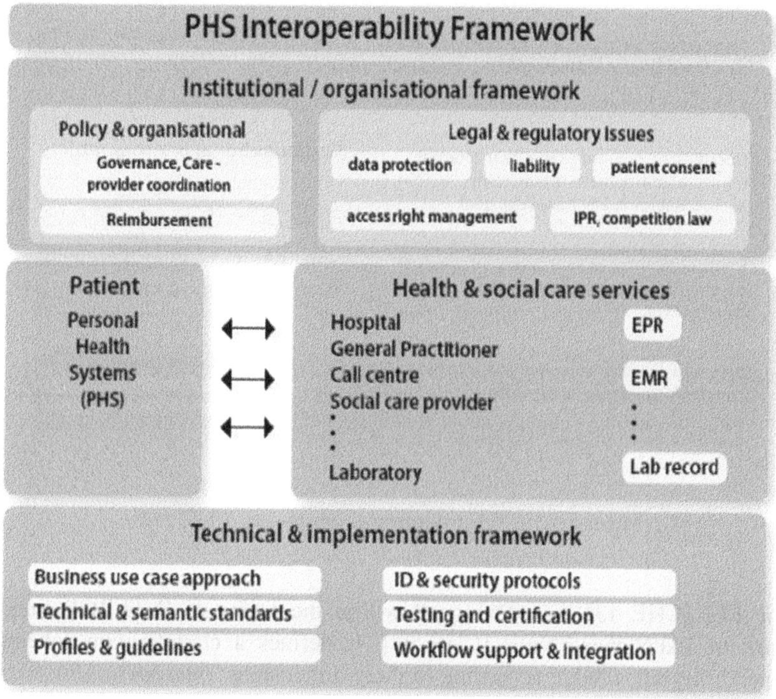

Figure 9.9. *PHS Interoperability framework [STR 11]*

9.5. Discussion

In this section, we compare the interoperability frameworks in the health domain based on the interoperability dimensions defined by the European Interoperability Framework (EIF) and the Framework of Enterprise Interoperability (FEI). Our choice is motivated by the generic nature of EIF and the fact that FEI is defined within the general perspective of an enterprise as a system, where a health organization can also be considered as an enterprise.

The EIF defines four levels of interoperability taking into account the political context: legal interoperability, organizational interoperability, semantic interoperability and technical interoperability, as depicted in Figure 9.7.

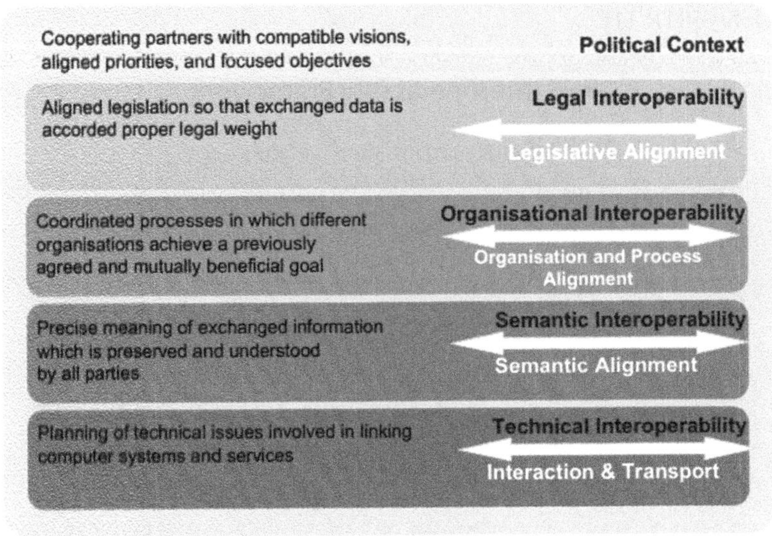

Figure 9.10. *EIF Interoperability levels [EUR 11]*

The FEI [CHE 13] was developed within the frame of INTEROP European Network of Excellence (NoE) [CHE 07]. It defines a classification scheme for interoperability knowledge according to three dimensions: *interoperability barriers*, *interoperability approaches* and enterprise *interoperability concerns* (Figure 9.11).

It is worth noting that interoperability barriers defined by FEI are based on interoperability levels, defined by EIF. Hence, FEI interoperability barriers will be omitted in this comparative analysis to avoid repetitions.

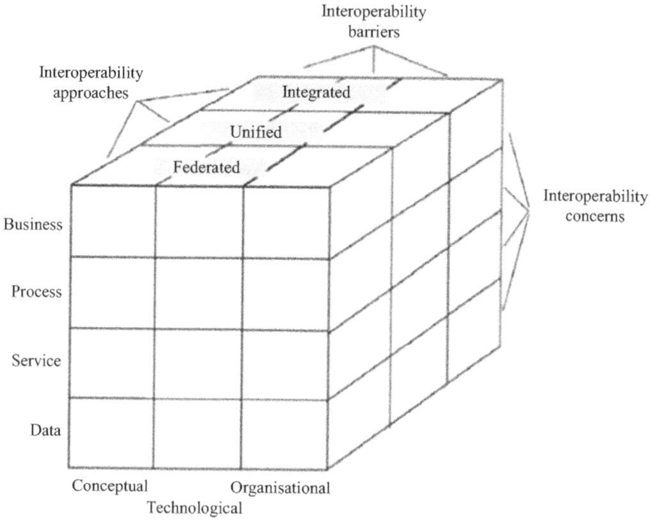

Figure 9.11. *FEI Interoperability dimensions*

Before comparing the reviewed health frameworks with criteria from EIF and FEI, it is important to give a general overview of the domain covered by these frameworks. This is given in Table 9.2.

	eHealth EIF	HIS IF	eHealth IF	PHS IF
Interoperability dimensions	– Legal – Organizational – Technical – Semantic	– Semantic – Technical	– Organizational – Technical – Informational	– Legal – Organizational – Technical
Case study orientation	Yes	–	Yes	Yes

Table 9.2. *Coverage of health interoperability frameworks*

Comparison criteria are based on the four interoperability levels of EIF and two dimensions of the general-framework FEI. Indications alongside each criterion in each table denote the corresponding coverage in each particular eHealth interoperability framework:

– √ indicates that the studied framework has adopted an approach for this criterion, without judging whether this approach provides full or partial coverage for the issue;

– × refers to the lack of a tangible approach to this issue;

– ? characterizes a criterion when the information gathered by the studied framework specifications is inadequate to evaluate this criterion.

9.5.1. *Interoperability levels*

EIF defines four levels of interoperability. Each deserves special attention when a new public service is established: *Legal interoperability* is concerned with relevant legislation relating to data exchange, including data protection legislation, when seeking to establish a European public service. *Organizational interoperability* is concerned with how organizations, such as public administrations in different Member States, cooperate to achieve their mutually agreed goals. *Semantic interoperability* ensures that the precise meaning of exchanged information is understood and preserved throughout exchanges between parties. *Technical interoperability* covers the technical aspects of linking information systems. It includes aspects such as interface specifications, interconnection services, data presentation and exchange.

Table 9.3 illustrates the coverage of the reviewed frameworks toward these four levels.

	eHealth EIF	HIS IF	eHealth IF	PHS IF
Legal	√	?	?	√
Semantic	√	√	√	√
Organizational	√	×	√	√
Technical	√	√	√	√

Table 9.3. *Coverage of interoperability levels*

9.5.2. *Interoperability concerns*

They represent the areas concerned by interoperability in an enterprise. Four concerns are defined, namely *business interoperability* (work in a harmonized way to share and develop business between companies despite the difference of methods, decision-making, culture of enterprises, etc.), *process interoperability* (make various processes work together; in the interworked enterprise, the aim will be to connect internal processes of two companies to create a common process), *service interoperability* (making various services or applications work together by solving the syntactic and semantic differences) and *data interoperability* (making different data models with different query languages work together to share information obtained from heterogeneous systems).

Table 9.4 illustrates the coverage of the reviewed frameworks toward the above-mentioned four enterprise interoperability concerns.

	eHealth EIF	HIS IF	eHealth IF	PHS IF
Business	×	?	√	√
Process	?	√	√	√
Service	?	√	√	√
Data	√	√	√	√

Table 9.4. *Coverage of interoperability concerns*

9.5.3. *Interoperability approaches*

There are three basic ways to relate entities together to establish interoperations: *the integrated approach* (characterized by the existence of a common format for all the constituents systems), the *unified approach* (characterized by the existence of a common format, but at a meta-level) and the *federated approach* (in which no common format is defined). This approach maintains the identity of interoperating systems; nothing is imposed by one party or another and interoperability is managed in an ad hoc manner. Table 9.5 illustrates the coverage of the reviewed frameworks toward these interoperability approaches.

	eHealth EIF	HIS IF	eHealth IF	PHS IF
Integrated	×	×	×	×
Unified	×	×	×	×
Federated	×	×	×	×

Table 9.5. *Coverage of interoperability approaches*

9.5.4. *Discussion*

The review of the different aspects and the frameworks coverage with respect to the dimensions defined by the EIF and FEI enables us to identify the main elements in health interoperability that are taken into account by the existing frameworks and the elements that are not considered within these frameworks.

Legal, organizational, technical and semantic levels are the main relevant dimensions that we find in the health interoperability domain. Some of them are missed, as shown in Tables 9.3 and 9.4.

The comparison has clearly shown that none of the reviewed interoperability frameworks, in section 9.3, considers interoperability approaches and thus proposes a way to deal with a specific interoperability problem (see Table 9.5).

Suggesting a unified framework that takes into account the existing ones and integrates the required dimensions of interoperability would allow us to cover the whole domain of interoperability and go beyond the existing approaches.

In order to facilitate the design of this unified framework, we believe that incorporating usage scenarios of the electronic patient record in an extended health pathway, particularly in supporting cross-border healthcare, is a very effective way to identify the dimensions of the future framework. These scenarios would be especially useful when considering cases where the patient data are, partially, supplied by information from medical devices. This is the approach that we intend to adopt in our future work to define the dimensions of the unified framework.

9.6. Conclusion and future work

In the last decades, developing interoperable systems has never been more important in eHealth ecosystems to enhance open-information exchanges and sharing among several heterogeneous systems. Indeed, eHealth interoperability is identified as the main prerequisite to ensure the seamless exchange of health-related information among the various players involved in patient management.

As important as this is, interoperability in the healthcare field is still in its early stages of development and is rather difficult to fully implement in all required dimensions. We have shown some of the factors that make the problem of achieving interoperability so complex to address.

We have made an attempt to study the most common dimensions of the interoperability concept in the healthcare field, namely the technical, semantic and organizational dimensions, and to appraise their degree of maturity. The first conclusion lies in a lack of common understanding and compromise about these dimensions. The second underlined some discrepancies among these common dimensions, which are at different levels of maturity by considering different frameworks.

To strengthen our study and enhance a more universal understanding, we surveyed the main interoperability frameworks in the health domain. A comparative analysis of the considered frameworks was then proposed on the basis of interoperability levels, as defined by the European Interoperability Framework (EIF), and interoperability concerns and approaches, as defined by the Framework of Enterprise Interoperability (FEI). This comparison enabled us to clearly identify the missing interoperability aspects in the existing frameworks within the health domain.

Our future work consists of organizing the dimensions and identifying their common features in order to propose a reconciliation between them. It could produce a new integrated framework that would be particularly well balanced for eHealth applications and would allow us to go beyond the existing frameworks which partially cover the domain or are not practical to use.

9.7. Bibliography

[AHM 11] AHMADIAN L., VAN ENGEN-VERHEUL M., BAKHSHI-RAIEZ F. *et al.*, "The role of standardized data and terminological systems in computerized clinical decision support systems: literature review and survey", *International Journal of Medical Informatics*, vol. 80, pp. 81–93, 2011.

[ASI 10] ASIP Santé-PRAS, Health Information Systems Interoperability Framework – Introduction Document, available at: http://esante.gouv.fr/sites/default/files/HIS_Interoperability_framework_v1%200%201_R.pdf, 2010.

[BEN 12] BENSON T., *Principles of Health Interoperability HL7 and SNOMED*, Springer Science & Business Media, 2012.

[BLA 11] BLACK A.D., CAR J., PAGLIARI C. *et al.*, "The impact of ehealth on the quality and safety of health care: a systematic overview", *PLoS Medicine*, vol. 8, e1000387, 2011.

[BOU 14] BOUX O., Interopérabilité et circuit du médicament: Le standard PN13 – SIPh2 v3 – Manuel de référence, 2nd edition, Phast, available at: http://www.departement-information-medicale.com/wp-content/uploads/2014/04/standard-PN13-SIPh_edition-2_couv.jpg, 2014.

[BRA 12] BRAA J., SAHAY S., *Integrated Health Information Architecture: Power to the Users: Design, Development, and Use*, Matrix Publishers, 2012.

[CAT 11] CATERINA M., HAYAT M., GUILLOREL M., Etude sur les technologies de l'information au service des nouvelles organisations de soins, available at: http://www.fafiec.fr/images/contenu/menuhaut/observatoire/etudes/2011/OPIIEC_%C3%A9tude_t%C3%A9l%C3%A9m%C3%A9decine_Vdef.pdf, 2011.

[CEN 05a] CEN/ISSS, Current and future standardization issues in the eHealth domain: Achieving interoperability – Part 2, CEN/ISSS eHealth Standardization Focus Group, 2005.

[CEN 05b] CEN/ISSS, Current and future standardization issues in the eHealth domain: Achieving interoperability – Part 1, CEN/ISSS eHealth Standardization Focus Group, 2005.

[CHE 07] CHEN D., DASSISTI M., ELVEAETER B., Enterprise interoperability framework and knowledge corpus, Final report, Interop Deliv. DI 3, 2007.

[CHE 13] CHEN D., "Framework for enterprise interoperability and maturity model (CEN/ISO 11354)", in *Interoperability for Enterprise Software and Applications*, ISTE Ltd, London and John Wiley & Sons, New York, 2013.

[CHU 10] CHUVAKIN A., "The complete guide to log and event management", available at: http://www.novell.com/docrep/2010/03/Log_Event_Mgmt_WP_DrAntonChuvakin_March2010_Single_en.pdf (accessed September 6th 2017), 2010.

[COC 15] COCIR, eHealth Toolkit 2015 Integrated Care: Breaking the Silos, available at: http://www.cocir.org/uploads/media/15013.COC_2.pdf, 2015.

[EHE 14] eHealth Initiative, 2014 Results from Survey on Health Data Exchange, available at: https://www.ehidc.org/resources/476-2014-results-from-survey-on-health-data-exchange, 2014.

[EUR 11] European Commission, Directorate-General for Informatics, European interoperability framework (EIF): towards Interoperability for European public services, Publications Office, Luxembourg, 2011.

[EUR 13] European Commission, Directorate-General for the Information Society and Media, eHealth European Interoperability Framework vision on eHealth EIF, Deloitte & Touche, Publications Office, Luxembourg, 2013.

[EYS 01] EYSENBACH G., "What is e-health?", *Journal of Medical Internet Research*, vol. 3, 2001.

[GIB 07] GIBBONS P., ARZT N., FLEWELLING T. *et al.*, Coming to Terms: Scoping Interoperability for Health Care Health Level Seven EHR Interoperability Work Group, available at: https://www.hln.com/assets/pdf/Coming-to-Terms-February-2007.pdf, 2007.

[HEU 06] Heubusch K., "Interoperability: what it means, why it matters", *Journal of AHIMA*, vol. 77, pp. 26–30, 2006.

[I2H 05] I2-Health, Interoperability Definitions, i2-Health Interoperability Initiative for a European eHealth area [WWW Document], available at: http://www.i2-health.eu/wiki-container/InteroperabilityDefinitions (accessed 22 January 2015), 2005.

[IDA 08] IDABC, Draft document as basis for EIF 2.0, European Communities, 2008.

[IHT 14] IHTSDO, SNOMED CT Starter Guide, available at: https://confluence.ihtsdotools.org/display/DOCSTART/SNOMED+CT+Starter+Guide?preview=/28742871/29953149/doc_StarterGuide_Current-en-US_INT_20160817.pdf, 2014.

[JAY 88] JAY F., GOETZ J., "IEEE standard dictionary of electrical and electronics terms", *Institute of Electrical and Electronics Engineers*, 1988.

[KUB 11] KUBICEK H., CIMANDER R., SCHOLL H.J., "Layers of interoperability", in *Organizational Interoperability in E-Government*, Springer, 2011.

[NEH 07] NEHTA, Interoperability Framework – Version 2.0, available at: https://www.digitalhealth.gov.au/implementation-resources/ehealth-foundations/EP-1144-2007/NEHTA-1146-2007, 2007.

[PAG 05] PAGLIARI C., SLOAN D., GREGOR P. *et al.*, "What is eHealth (4): a scoping exercise to map the field", *Journal of Medical Internet Research*, vol. 7, e9, 2005.

[PIA 09] PIANYKH O.S., *Digital Imaging and Communications in Medicine (DICOM): A Practical Introduction and Survival Guide*, Springer Science & Business Media, 2009.

[PIN 09] PINGAUD H., "Rationalité du développement de l'interopérabilité dans les organisations", in *Management Des Technologies Organisationnelles*, Presses des MINES, 2009.

[REY 09] REYNOLDS M., EVANGELIDIS P., HEIDENREICH G. *et al.*, eHealth-INTEROP Report in response to eHealth Interoperability Standards Mandate (No. SA/CEN/ENTR/000/2007-20), CEN/CENELEC/ETSI, 2009.

[RHO 10] RHOADS J.G., COOPER T., FUCHS K. *et al.*, "Medical device interoperability and the Integrating the Healthcare Enterprise (IHE) initiative", *Biomedical Instrumentation & Technology*, available at: https://s3.amazonaws.com/rdcms-aami/files/production/public/FileDownloads/HT_Interoperability/2010.ith.Interoperability_Rhoads.pdf, pp. 21–7, 2010.

[RIA 07] RIALLE V., Technologies nouvelles susceptibles d'améliorer les pratiques gérontologiques et la vie quotidienne des malades âgés et de leur famille, Ministère de la santé et des solidarités, 2007.

[ROT 93] ROTHSTEIN B., BOCQUENTIN J., "Implementing ASTM 1238 on a national scale in France: the H.PR.I.M. experience 1990–1992", *Studies in Health Technology and Informatics*, vol. 6, pp. 162–166, 1993.

[STR 11] STROETMANN V., THIEL R., WILSON P. *et al.*, Enabling Smart Integrated Care: Recommendations for Fostering Greater Interoperability of Personal Health Systems, Office for Official Publication of the European Communities, available at: https://joinup.ec.europa.eu/sites/default/files/files_epractice/sites/Enabling%20Smart%20Integrated%20Care%20-%20Recommendations%20for%20Fostering%20Greater%20Interoperability%20of%20Personal%20Health%20Systems.pdf, 2011.

[VER 10] VERNADAT F.B., "Technical, semantic and organizational issues of enterprise interoperability and networking", *Annual Reviews in Control*, vol. 34, pp. 139–144, 2010.

9.8. Glossary

ATC:	Anatomical Therapeutic Chemical
CDSS:	Clinical Decision Support Systems
COSTART:	Coding Symbols for a Thesaurus of Adverse Reaction Terms
CTS2:	Common Terminology Services 2
DeCS:	Health Sciences Descriptors (DeCS for its capital letters in Spanish)
FMA:	Foundational Model of Anatomy
GO:	Gene Ontology
ICD-10:	International Classification of Diseases 10th revision
ICPC-2:	International Classification of Primary Care 2nd edition
IHTSDO:	International Health Terminology Standards Development Organisation
IS:	Information Systems
LOINC:	Logical Observation Identifiers Names and Codes
MedDRA:	Medical dictionary for Regulatory Activities
MeSH:	Medical Subject Headings
SERAM:	Catalogue of the Radiology Spanish Society (SERAM for its capital letters in Spanish)
SNOMED CT:	Systematized Nomenclature of Medicine – Clinical Terms
UCUM:	Unified Code for Units of Measure

List of Authors

Bernard ARCHIMÈDE
École Nationale d'Ingénieurs de Tarbes
France

Jordi AYZA GRAELLS
Blocs TecnoCampus
Universitat Politècnica de Catalunya
Barcelona
Spain

Alain BERNARD
Ecole Centrale de Nantes
France

Séverine BLANC SERRIER
University of Bordeaux
IMS
France

Yacine BOUALLOUCHE
Ecole Centrale de Nantes
France

Jean-Paul BOURRIÈRES
University of Bordeaux
IMS
France

Mamadou Samba CAMARA
Cheikh Anta Diop University
Dakar
Senegal

Vincent CHAPURLAT
Ecole des Mines d'Alès
France

Aurélie CHARLES
University Lyon 2
DISP
France

David CHEN
University of Bordeaux
IMS
France

Raphaël CHENOUARD
LUNAM University
Ecole Centrale de Nantes
France

Catherine DA CUNHA
LUNAM University
Ecole Centrale de Nantes
France

Nicolas DACLIN
Ecole des Mines d'Alès
France

Guy DOUMEINGTS
University of Bordeaux
IMS
France
and
INTEROP Vlab
Brussels
Belgium

Yves DUCQ
University of Bordeaux
IMS
France

Rémy DUPAS
University of Bordeaux
IMS
France

Franck FONTANILI
University of Toulouse–Mines Albi
Albi
France

Wided GUÉDRIA
Luxembourg Institute of Science
and Technology
Alzette
Luxembourg

Léonard JANER-GARCÍA
Blocs TecnoCampus
Universitat Politècnica de Catalunya
Barcelona
Spain

Elyes LAMINE
University of Toulouse
Mines Albi- INU Champollion ISIS
Castres
France

Jacques LAMOTHE
University of Toulouse–Mines Albi
Albi
France

Matthieu LAURAS
University of Toulouse–Mines Albi
Albi
Francc

Yacine OUZROUT
University Lyon 2
DISP
France

Hervé PINGAUD
CNRS, LGC
University of Toulouse
France

Ariadna RIUS SOLER
TicSalut Foundation
Ministry of Health
Government of Catalonia
Barcelona
Spain

Gérald SANTUCCI
Directorate-General for
Communications Networks, Content
and Technology
European Commission
Brussels
Belgium

Sébastien TRUPTIL
University of Toulouse–Mines Albi
Albi
France

Bruno VALLESPIR
University of Bordeaux
IMS
France

Index

Other titles from

iSTE

in

Systems and Industrial Engineering – Robotics

2017

FEYEL Philippe
Robust Control, Optimization with Metaheuristics

MARÉ Jean-Charles
Aerospace Actuators 2: Signal-by-Wire and Power-by-Wire

POPESCU Dumitru, AMIRA Gharbi, STEFANOIU Dan, BORNE Pierre
Process Control Design for Industrial Applications

RÉVEILLAC Jean-Michel
Modeling and Simulation of Logistics Flows 1: Theory and Fundamentals
Modeling and Simulation of Logistics Flows 2: Dashboards, Traffic Planning and Management
Modeling and Simulation of Logistics Flows 3: Discrete and Continuous Flows in 2D/3D

2016

ANDRÉ Michel, SAMARAS Zissis
Energy and Environment
(Research for Innovative Transports Set - Volume 1)

AUBRY Jean-François, BRINZEI Nicolae, MAZOUNI Mohammed-Habib
Systems Dependability Assessment: Benefits of Petri Net Models (Systems Dependability Assessment Set - Volume 1)

BLANQUART Corinne, CLAUSEN Uwe, JACOB Bernard
Towards Innovative Freight and Logistics (Research for Innovative Transports Set - Volume 2)

COHEN Simon, YANNIS George
Traffic Management (Research for Innovative Transports Set - Volume 3)

MARÉ Jean-Charles
Aerospace Actuators 1: Needs, Reliability and Hydraulic Power Solutions

REZG Nidhal, HAJEJ Zied, BOSCHIAN-CAMPANER Valerio
Production and Maintenance Optimization Problems: Logistic Constraints and Leasing Warranty Services

TORRENTI Jean-Michel, LA TORRE Francesca
Materials and Infrastructures 1 (Research for Innovative Transports Set - Volume 5A)
Materials and Infrastructures 2 (Research for Innovative Transports Set - Volume 5B)

WEBER Philippe, SIMON Christophe
Benefits of Bayesian Network Models
(Systems Dependability Assessment Set – Volume 2)

YANNIS George, COHEN Simon
Traffic Safety (Research for Innovative Transports Set - Volume 4)

2015

AUBRY Jean-François, BRINZEI Nicolae
Systems Dependability Assessment: Modeling with Graphs and Finite State Automata

BOULANGER Jean-Louis
CENELEC 50128 and IEC 62279 Standards

BRIFFAUT Jean-Pierre
E-Enabled Operations Management

MISSIKOFF Michele, CANDUCCI Massimo, MAIDEN Neil
Enterprise Innovation

2014

CHETTO Maryline
Real-time Systems Scheduling
Volume 1 – Fundamentals
Volume 2 – Focuses

DAVIM J. Paulo
Machinability of Advanced Materials

ESTAMPE Dominique
Supply Chain Performance and Evaluation Models

FAVRE Bernard
Introduction to Sustainable Transports

GAUTHIER Michaël, ANDREFF Nicolas, DOMBRE Etienne
Intracorporeal Robotics: From Milliscale to Nanoscale

MICOUIN Patrice
Model Based Systems Engineering: Fundamentals and Methods

MILLOT Patrick
Designing Human–Machine Cooperation Systems

NI Zhenjiang, PACORET Céline, BENOSMAN Ryad, REGNIER Stéphane
Haptic Feedback Teleoperation of Optical Tweezers

OUSTALOUP Alain
Diversity and Non-integer Differentiation for System Dynamics

REZG Nidhal, DELLAGI Sofien, KHATAD Abdelhakim
Joint Optimization of Maintenance and Production Policies

STEFANOIU Dan, BORNE Pierre, POPESCU Dumitru, FILIP Florin Gh.,
EL KAMEL Abdelkader
Optimization in Engineering Sciences: Metaheuristics, Stochastic Methods and Decision Support

2013

ALAZARD Daniel
Reverse Engineering in Control Design

ARIOUI Hichem, NEHAOUA Lamri
Driving Simulation

CHADLI Mohammed, COPPIER Hervé
Command-control for Real-time Systems

DAAFOUZ Jamal, TARBOURIECH Sophie, SIGALOTTI Mario
Hybrid Systems with Constraints

FEYEL Philippe
Loop-shaping Robust Control

FLAUS Jean-Marie
Risk Analysis: Socio-technical and Industrial Systems

FRIBOURG Laurent, SOULAT Romain
Control of Switching Systems by Invariance Analysis: Application to Power Electronics

GROSSARD Mathieu, REGNIER Stéphane, CHAILLET Nicolas
Flexible Robotics: Applications to Multiscale Manipulations

GRUNN Emmanuel, PHAM Anh Tuan
Modeling of Complex Systems: Application to Aeronautical Dynamics

HABIB Maki K., DAVIM J. Paulo
Interdisciplinary Mechatronics: Engineering Science and Research Development

HAMMADI Slim, KSOURI Mekki
Multimodal Transport Systems

JARBOUI Bassem, SIARRY Patrick, TEGHEM Jacques
Metaheuristics for Production Scheduling

KIRILLOV Oleg N., PELINOVSKY Dmitry E.
Nonlinear Physical Systems

LE Vu Tuan Hieu, STOICA Cristina, ALAMO Teodoro, CAMACHO Eduardo F., DUMUR Didier
Zonotopes: From Guaranteed State-estimation to Control

MACHADO Carolina, DAVIM J. Paulo
Management and Engineering Innovation

MORANA Joëlle
Sustainable Supply Chain Management

SANDOU Guillaume
Metaheuristic Optimization for the Design of Automatic Control Laws

STOICAN Florin, OLARU Sorin
Set-theoretic Fault Detection in Multisensor Systems

2012

AÏT-KADI Daoud, CHOUINARD Marc, MARCOTTE Suzanne, RIOPEL Diane
Sustainable Reverse Logistics Network: Engineering and Management

BORNE Pierre, POPESCU Dumitru, FILIP Florin G., STEFANOIU Dan
Optimization in Engineering Sciences: Exact Methods

CHADLI Mohammed, BORNE Pierre
Multiple Models Approach in Automation: Takagi-Sugeno Fuzzy Systems

DAVIM J.Paulo
Lasers in Manufacturing

DECLERCK Philippe
Discrete Event Systems in Dioid Algebra and Conventional Algebra

DOUMIATI Moustapha, CHARARA Ali, VICTORINO Alessandro, LECHNER Daniel
Vehicle Dynamics Estimation using Kalman Filtering: Experimental Validation

GUERRERO José A, LOZANO Rogelio
Flight Formation Control

HAMMADI Slim, KSOURI Mekki
Advanced Mobility and Transport Engineering

MAILLARD Pierre
Competitive Quality Strategies

MATTA Nada, VANDENBOOMGAERDE Yves, ARLAT Jean
Supervision and Safety of Complex Systems

POLER Raul *et al.*
Intelligent Non-hierarchical Manufacturing Networks

TROCCAZ Jocelyne
Medical Robotics

YALAOUI Alice, CHEHADE Hicham, YALAOUI Farouk, AMODEO Lionel
Optimization of Logistics

ZELM Martin *et al.*
Enterprise Interoperability –I-EASA12 Proceedings

2011

CANTOT Pascal, LUZEAUX Dominique
Simulation and Modeling of Systems of Systems

DAVIM J. Paulo
Mechatronics

DAVIM J. Paulo
Wood Machining

GROUS Ammar
Applied Metrology for Manufacturing Engineering

KOLSKI Christophe
Human–Computer Interactions in Transport

LUZEAUX Dominique, RUAULT Jean-René, WIPPLER Jean-Luc
Complex Systems and Systems of Systems Engineering

ZELM Martin, *et al.*
Enterprise Interoperability: IWEI2011 Proceedings

2010

BOTTA-GENOULAZ Valérie, CAMPAGNE Jean-Pierre, LLERENA Daniel, PELLEGRIN Claude
Supply Chain Performance / Collaboration, Alignement and Coordination

BOURLÈS Henri, GODFREY K.C. Kwan
Linear Systems

BOURRIERES Jean-Paul
Proceedings of CEISIE'09

CHAILLET Nicolas, REGNIER Stéphane
Microrobotics for Micromanipulation

DAVIM J. Paulo
Sustainable Manufacturing

GIORDANO Max, MATHIEU Luc, VILLENEUVE François
Product Life-Cycle Management / Geometric Variations

LOZANO Rogelio
Unmanned Aerial Vehicles / Embedded Control

LUZEAUX Dominique, RUAULT Jean-René
Systems of Systems

VILLENEUVE François, MATHIEU Luc
Geometric Tolerancing of Products

2009

DIAZ Michel
Petri Nets / Fundamental Models, Verification and Applications

OZEL Tugrul, DAVIM J. Paulo
Intelligent Machining

PITRAT Jacques
Artificial Beings

2008

ARTIGUES Christian, DEMASSEY Sophie, NERON Emmanuel
Resources–Constrained Project Scheduling

BILLAUT Jean-Charles, MOUKRIM Aziz, SANLAVILLE Eric
Flexibility and Robustness in Scheduling

DOCHAIN Denis
Bioprocess Control

LOPEZ Pierre, ROUBELLAT François
Production Scheduling

THIERRY Caroline, THOMAS André, BEL Gérard
Supply Chain Simulation and Management

2007

DE LARMINAT Philippe
Analysis and Control of Linear Systems

DOMBRE Etienne, KHALIL Wisama
Robot Manipulators

LAMNABHI Françoise *et al.*
Taming Heterogeneity and Complexity of Embedded Control

LIMNIOS Nikolaos
Fault Trees

2006

FRENCH COLLEGE OF METROLOGY
Metrology in Industry

NAJIM Kaddour
Control of Continuous Linear Systems

Printed and bound by CPI Group (UK) Ltd, Croydon, CR0 4YY
16/09/2026
14963072-0002